Practical Design Considerations for Isolated DC-DC Converters

Alexander Isurin • Mark J. Scott

Practical Design Considerations for Isolated DC-DC Converters

Springer

Alexander Isurin
Power Converters Future, LLC
Dublin, OH, USA

Mark J. Scott
Electrical and Computer Engineering
Miami University
Oxford, OH, USA

ISBN 978-3-031-94426-0 ISBN 978-3-031-94427-7 (eBook)
https://doi.org/10.1007/978-3-031-94427-7

This Springer imprint is published by the registered company Springer Nature Switzerland AG
The registered company address is: Gewerbestrasse 11, 6330 Cham, Switzerland

I dedicate this book to my wife, Ludmila, for her love and support throughout the process of our work on this book, and to the United States that gave me a chance to fulfill my professional goals and leave my legacy in the form of my patents and this manuscript.

—Alexander Isurin

This book is dedicated to God, for the challenges and opportunities that have shaped me; to my parents, Mark and Darlene Scott, for a lifetime of support and guidance; and to my wife, strongest advocate, and best friend, Kendra Ellinger, for the sacrifices she has made to help me chase my dreams.

—Mark J. Scott

Acknowledgments

First, we want to acknowledge Alec Cook for all of his support and assistance during the past 25 years. Additionally, we want to thank Havis, Inc. for allowing us to include their images and data in our book and Alec Cook for coordinating this effort.

We would also like to acknowledge the following students: Tra Yen Nhu Phan for her help in creating printed circuit boards and setting up experiments; James Nippa and Logan Rauh for creating images and tables; and Aaliyah Garrett for assisting with the Alt Text preparation.

We want to thank ECCE 2017 and ECCE 2022 for allowing us to present these ideas at their conferences and for facilitating a venue to dialogue with our professional network.

Portions of this book are based upon work supported in part by the National Institute for Standards and Technology under Award Number 70NANB22H007. Neither the United States Government nor any agency thereof, nor any of their employees, makes any warranty, express or implied, or assumes any legal liability or responsibility for the accuracy, completeness, or usefulness of any information, apparatus, product, or process disclosed, or represents that its use would not infringe privately owned rights.

The content of this book is the original, creative work of the authors. We utilized OpenAI's ChatGPT to check grammar and spelling and improve the clarity of portions of the text.

Competing Interests The authors have no competing interests to declare that are relevant to the content of this manuscript.

Contents

Abbreviations

Symbols

BOM	Bill of materials
EMC	Electromagnetic compatibility
EMI	Electromagnetic interference
GaN	Gallium nitride
IGBT	Insulated gate bipolar transistor
MOSFET	Metal-oxide-semiconductor field-effect transistor
R&D	Research and development
PCB	Printed circuit board
Si	Silicon
SiC	Silicon carbide
SMA	Surface-mount assembly
SMPS	Switched-mode power supply
WBG	Wide bandgap
ZCS	Zero-current switching
ZVS	Zero-voltage switching

Variables

C_{DG}	Miller capacitance
C_{DIE}	Body capacitance of a transistor
C_{GS}	Gate-source capacitance
C_{ISS}	Input capacitance
C_{OSS}	Output capacitance
C_{RSS}	Reverse transfer capacitance
Co/XFMR	Coefficient of power transformer use
dV/dt	Voltage slew rate

dI/dt	Current slew rate
f_{COM}	Commutation frequency
L_{CS}	Common-source inductance
L_{leak}	Leakage inductance
L_{mag}	Magnetizing inductance
L_{WB}	Inductance of the wire bonds
$P–C$	Duty cycle of power conversion
P_D	Power dissipation in a semiconductor device
R_{DS_ON}	On-resistance of a MOSFET
R_I	Internal gate resistor of a transistor
$R_{th,JC}$	Thermal resistance from the semiconductor junction to the device's case
T_J	Junction temperature of a semiconductor die

List of Figures

List of Tables

Chapter 1
Introduction

The motivation to write this book arises from a series of pivotal moments in the professional journey of the authors, each highlighting the need for a dedicated resource on designing high-current, cost-effective, isolated DC–DC converters. For several decades, the authors have been deeply involved in exploring and addressing practical challenges in power conversion, and this book represents the culmination of those efforts. The text is to serve as a resource for trained engineers and is not intended to be a traditional textbook or handbook. Instead, it focuses on real-world scenarios and offers engineers a roadmap to navigate through the challenges of a cost-effective design. We prioritize practical applications, component selection, and design trade-offs. This book presents the authors' philosophy that real-world engineering is as much about problem-solving as it is about theory. The authors hope this book will not only provide answers, but will also inspire engineers to ask better questions and approach their work with renewed confidence.

1.1 Evolution of the Book

In 2014, one of the authors received the *Best Paper Award* at the *2014 IEEE Transportation Electrification Conference and Exposition* in Detroit, MI [1]. The award-winning paper, *Philosophy of Topology and Components Selection for Cost and Performance Converters*, provided a framework for understanding how to balance cost and performance in converter design. This recognition underscored the importance of creating methodologies that address real-world challenges rather than purely theoretical ones.

Building on this success, the authors presented a tutorial titled *Design Guidelines for Isolated DC–DC Converters Selection Based on Cost, Reliability, and Compatibility* at the *2017 IEEE Energy Conversion Congress and Exposition* [2]. This tutorial focused on equipping engineers with the tools to make informed design

A. Isurin, M. J. Scott, *Practical Design Considerations for Isolated DC-DC Converters*, https://doi.org/10.1007/978-3-031-94427-7_1

choices, particularly in scenarios where cost constraints and reliability requirements intersect. The positive reception and engagement during this session reaffirmed the need for more accessible, practical knowledge in this area.

In 2022, the authors presented another tutorial, *Design Strategies for High-Power, High-Current, Isolated DC/DC Converters Using Soft-Switching Technology and Silicon Carbide Transistors*, at the *2022 IEEE Energy Conversion Congress and Exposition* [3]. This presentation delved into advanced topics such as the use of silicon carbide (SiC) transistors and soft-switching techniques, addressing the growing demand for efficient, high-power converters in modern applications. During these events, the authors engaged in extensive discussions with attendees, many of whom were experienced engineers.

From these interactions—both during the tutorials and through follow-up conversations via email—the authors noticed a consistent trend. Most questions were centered on practical implementation: how to select components, optimize topologies, and address design trade-offs in real-world scenarios. Theoretical inquiries were relatively rare, emphasizing the need for actionable guidance tailored to solving practical problems.

One area of particular interest was high-current, low-voltage converters, specifically those handling currents exceeding hundreds of amperes and operating below 60 V. While this is a specialized niche, its importance is growing, especially in sectors like transportation, where cost, reliability, and efficiency are critical. The transportation industry, in particular, is undergoing rapid electrification, and the demand for scalable, cost-sensitive solutions is increasing.

The authors recognize a significant gap in the resources available to engineers working in this domain. Most existing books and guides focus on either broad theoretical principles or specific technologies but fail to provide a comprehensive methodology for addressing the unique challenges of high-current, low-voltage applications. Moreover, the authors realize that while no single book can answer all the questions that arise during the development process, a resource offering a structured approach to problem-solving can empower engineers to find solutions on their own.

The authors believe that providing a clear methodology—rooted in both industry experience and academic research—can significantly assist design engineers in tackling the complex questions that emerge during development. This book seeks to bridge the gap by offering a blend of practical insights, design philosophies, and actionable tools derived from the authors' extensive experience.

Thus, this book was born. Its focus is on practical directions and tools for developing isolated DC–DC converters with output power ranging from 2 to 4 kW and higher power levels as well as high current levels (over 100–200 A) on the low-voltage side. It is designed to serve as a resource for trained engineers but is also accessible to graduate students completing their studies.

This book is not intended to be a traditional textbook or handbook. Instead, it focuses on practical implementation, offering engineers a roadmap for navigating the challenges of cost-effective design. The content prioritizes practical applications, component selection, and design trade-offs, reflecting the authors' philosophy that real-world engineering is as much about problem-solving as it is about theory.

The authors are confident that the principles, methodologies, and case studies presented in this book will provide valuable guidance to engineers. However, they also acknowledge that no resource can replace the critical thinking and ingenuity required of design engineers. Each new project presents unique challenges, and the ability to adapt and innovate is paramount.

In this sense, the authors hope this book will not only provide answers but also inspire engineers to ask better questions and approach their work with renewed confidence. Ultimately, the success of this book will be determined by its readers, and the authors sincerely hope it proves to be a helpful and practical resource for anyone working in this dynamic and evolving field.

1.2 Description of the Book's Contents

The rest of the book is structured as follows. Chapters 2–5 provide background and context for the remaining six chapters in the book, where we compare six isolated DC–DC converter topologies, describe power magnetics, talk about advanced topologies, present a case study, and finish the book with our closing remarks.

Chapter 2 presents the authors' core engineering philosophy: the primary goal of engineering is to generate profits for companies. This chapter explores how our philosophy influences the design process and provides strategies for design engineers to develop cost-effective converters that are reliable and compatible with the systems they interact with while simultaneously meeting customer specifications. A key takeaway from this section is that Efficiency should not be treated as the ultimate goal but rather as a variable to be optimized alongside cost.

Chapter 3 builds on the philosophical discussion in Chap. 2. It encourages the reader to reflect on the characteristics of ideal power conversion topologies—100% efficiency and continuous power flow from the input to the output of the converter. While acknowledging this as an impossible achievement, we show that isolated DC–DC converters are the optimal choice for our applications of interest. Design engineers have primarily two approaches for utilizing semiconductor devices: hard-switching and soft-switching. This chapter explains why soft-switching is the preferred approach. Additionally, it examines the modes of operation for power converters: constant voltage, constant current, and constant power modes. Finally, it explores dosed energy transfer, the mechanism employed in resonant converters.

In Chap. 4, we discuss semiconductor switching devices. Currently, power converters primarily use silicon (Si) MOSFETs and IGBTs in all but a few niche applications. At the time of this writing, wide bandgap (WBG) power devices are becoming reliable and competitively priced compared to traditional semiconductor technologies. We briefly introduce WBG technology, as manufacturers are beginning to incorporate these components into their products.

A significant portion of Chap. 4 focuses on comparing the advantages and disadvantages of Si MOSFETs, SiC MOSFETs, and SiC cascode devices. We encourage readers to understand these points before adopting WBG components. For instance, while SiC components offer lower specific on-resistances, their smaller dies pose challenges for thermal management. Additionally, the fast-switching speeds of WBG devices necessitate careful consideration of how parasitic inductances impact long-term reliability. The key takeaway is that **wide bandgap devices must be used with soft-switching technology** to ensure the long-term reliability of a product.

Chapter 5 introduces the reader to the many design considerations engineers must understand to produce reliable converters. Then, it builds on Chaps. 3 and 4, describing how to control semiconductor switching devices so that power converters meet design objectives. The chapter begins with regulation techniques for power converters, including fixed-frequency and variable-frequency operations. It then reviews several algorithms for implementing pulse-width modulation (PWM) and variable-frequency modulation. Next, the chapter explains how to select the commutation frequency for the chosen modulation strategy. Finally, it examines how regulation is performed, at the system level, in isolated DC–DC converters.

Chapter 6 is the heart of the book. First, we introduce a novel set of metrics that provide design engineers with a systematic method for comparing power electronic topologies in a given application. These metrics offer a unique framework for evaluating design trade-offs and will be valuable in assessing topologies across a wide range of power conversion applications.

Second, we use our metrics to evaluate six base-level isolated DC–DC converters that are capable of (1) a high conversion ratio, (2) current levels exceeding 100–200 A on the low-voltage side, and (3) output powers ranging from 2 to 4 kW or higher. These requirements are also widely used in the transportation sector, which is the case study explored in Chap. 6, but also apply to other areas like telecommunications.

Obviously, it is impossible to comprehensively address multiple applications within a single book. Therefore, the goal of Chap. 6 is to provide an archetype for design engineers to evaluate solutions for their design criteria. Additionally, we do not explore the many modifications that exist for these base-level topologies due to the breadth of the topic.

We wish to draw our readers' attention to a major point of our philosophy: **the design engineer must focus their attention on addressing the root causes of a topology's weaknesses rather than fighting against the consequences of those weaknesses**. This philosophy is not exclusive to power electronics; it is relevant to all engineering. Addressing the symptoms of a converter's shortcomings, rather than their sources, does not produce a cost-effective, reliable solution. Furthermore, topologies that require fewer modifications are generally better choices. However, Chap. 10 provides an example of how combining the strengths of multiple topologies can overcome many of their collective weaknesses. These innovations can expand a company's portfolio and offer a competitive advantage in specific design spaces.

Chapter 7 is the proverbial brain for the book as it analyzes the data generated from Chap. 6's investigation. Our analysis considers both step-up and step-down modes. First, the topologies are compared by the cost estimates presented in Chap. 6. The metrics are dollar-value estimates for the power stage, auxiliary circuitry, filters, and total system costs; these estimates are based on the current dollar-values in 2024. Next, it compares the operation areas of each of the topologies using the metrics defined in Chap. 6: the duty cycle of power conversion, the coefficient using power transformer, the stability conversion, the range of the soft-switching, its ability to soft-start, its reliability, and its maximum commutation frequency. Finally, each topology is evaluated by its application area where we consider both its maximum potential and its cost-effective range.

Chapter 8 provides a high-level overview of magnetics, focusing on power transformers and techniques for designing high-current (100–200 A) magnetics. It compares planar transformers with integrated transformers, the latter playing a key role in this work. The chapter also explores methods for designing and utilizing power inductors on both high- and low-voltage sides of a converter. It concludes with procedures for measuring non-ideal parameters of magnetic components, including leakage inductance, parasitic resistance, and transformer losses.

Chapter 9 reviews the practical implementation and operation of isolation DC–DC converters in bidirectional DC–DC converters, multiport converters, and multilevel configurations. The main takeaway from this section is that these converters are highly individualized and depend on specific technical specifications. The design engineer has numerous considerations when realizing this hardware.

Chapter 10 demonstrates how the authors applied the principles in the preceding chapters as part of a recent project. Our design target was a converter capable of interfacing a low-voltage DC bus with a high-voltage DC bus in a microgrid application. A significant challenge was achieving high efficiency across a wide operating range.

We evaluated the CLLC converter and the resonance with PWM converter for this application. Each presented its own set of weaknesses. By combining the strengths of these topologies, we developed an innovative new design. This chapter details the entire design process for the converter and concludes with experimental results. This portion of the book was supported by a research project with PowerAmerica and financial support from the National Institute for Standards and Technology under Award Number 70NANB22H007.

Chapter 11 concludes the book by exploring the future role of wide bandgap (WBG) devices in isolated DC–DC converters. It emphasizes the importance of soft-switching technology when designing converters with these advanced components. The chapter revisits the properties of an ideal converter in detail and concludes with final reflections on the topics discussed throughout the book. The authors included an appendix with supplementary information on measuring voltage and current.

1.3 Supplemental

The content of this chapter is the original, creative work of the authors. We utilized OpenAI's ChatGPT to check grammar and spelling and improve the clarity of portions of the text.

References

1. A. Isurin, A. Cook, Philosophy of topology and components selection for cost and performance in automotive converters, in *2014 IEEE Transportation Electrification Conference and Expo (ITEC), Dearborn, MI, USA*, (2014), pp. 1–5
2. A. Isurin, M.J. Scott, Using soft-switching technology to design high-power, high-current, isolated, DC/DC converters that achieve low-cost, high reliability, and electromagnetic compliance, in *2017 IEEE Energy Conversion Congress and Exposition (ECCE 2017), Cincinnati, OH, USA*, (2017)
3. A. Isurin, M.J. Scott, Design strategies for high-power, high-current, isolated DC/DC converters using soft-switching technology and silicon carbide transistors, in *2022 IEEE Energy Conversion Congress and Exposition (ECCE 2022), Detroit, MI, USA*, (2022)

Chapter 2
Engineering Philosophy

The authors aim to share their knowledge and experience in the development of new converters. Throughout the development process, several key engineering philosophy principles are emphasized, which serve as guidelines for design engineers in the power electronics field.

The primary objective of this book is to assist power electronics design engineers who are currently engaged or are preparing to work in the dynamic power electronics industry. This work represents a synthesis of extensive international experience and academic insight. We will explore the practical implementation of isolated DC–DC converters, typically rated at 2 kW and above, focusing on achieving effective and cost-efficient results. However, the principles of this engineering philosophy extend beyond power electronics applications to any engineering discipline.

2.1 Core Engineering Philosophy Principles

This philosophy is rooted in the following advancements:

1. **Technical status of a country**
   - The capabilities and innovations in power electronics often reflect the technical status of a country. Countries with strong educational systems, research institutions, and industrial capabilities generally lead in technological advancements.
2. **Benchmark industries**
   - The automotive and aviation industries serve as critical benchmarks for engineering standards. The challenges and requirements encountered in these two sectors provide insights that benefit other fields, including medical, industrial manufacturing, and telecommunications.

A. Isurin, M. J. Scott, *Practical Design Considerations for Isolated DC-DC Converters*, https://doi.org/10.1007/978-3-031-94427-7_2

Automotive industry

- There are three primary drivers in the automotive industry. The most important is *cost*, then *reliability*, and then *electromagnetic compatibility.*
  - **Cost**: The pressure to lower costs while maintaining quality is paramount, influencing design decisions and component selection.
  - **Reliability**: Automotive applications demand high reliability due to the safety-critical nature of vehicles.
  - **Electromagnetic compatibility**: As vehicles incorporate more electronic systems, ensuring compatibility is crucial to prevent interference.

Aviation industry

- The drivers in the aviation industry, from most important to least important, are *weight/power density*, *reliability*, *electromagnetic compatibility*, and *cost.*
  - **Weight/power density**: Aviation demands solutions with high power density that minimize weight, thereby maximizing fuel efficiency and load capacity.
  - **Reliability**: Similar to automotive applications, reliability is essential due to safety concerns and cost implications of failures.
  - **Electromagnetic compatibility**: Avoiding electromagnetic interference is critical in aviation to ensure the integrity of navigation and communication systems.
  - **Cost**: Cost efficiency remains important in both the production and operation of aircraft systems.

2.1.1 Principles of Power Electronics Engineering

Power electronics represent an application of physics to engineering, where effective engineering is a crucial tool for companies to achieve profitability. Engineering is fundamentally a business discipline; when executed effectively, it contributes to substantial profits, while poor engineering can lead to a range of issues, from unreliable operations to engineering disasters.

2.1.2 Total Cost Consideration

The total cost of a product and its development—including lifecycle and after-sale costs—must be the primary focus of engineering efforts. Engineers need to anticipate and avoid misguided design directions.

2.1.3 Design Process Emphasis

When developing power converters, it is considerably easier to make changes when the process remains theoretical, whether on paper or through simulations. This is why we emphasize understanding both the advantages and disadvantages of various topologies and components. Careful comparisons among different designs, components, and their suitable applications will enable engineers to make informed decisions.

2.1.4 Decision-Making Framework

Design engineers must follow a structured framework:

1. **Identify weaknesses**: Analyze the design to pinpoint potential pitfalls.
2. **Evaluate strengths**: Consider the benefits offered by various design choices.
3. **Make informed technical decisions**: Use the insights gained to arrive at a well-founded design approach.

Predictability in the behavior of topologies and components is crucial. During prototyping, unexpected challenges are likely to arise, and it is vital to minimize these surprises. A topology should not be viewed as a standalone unit; instead, engineers must consider the entire system, encompassing the input-to-output connections. This holistic view includes EMI filters, power transformers, and auxiliary circuits, all of which must work together seamlessly.

2.1.5 Conclusion

The authors believe that a good engineering book must include numerous practical examples of design, accompanied by thorough explanations of the reasoning behind those decisions. Although a power electronics design engineer must possess a deep understanding of power conversion, they must also be knowledgeable about economics. Every development process will encounter challenges, and it is the design engineer's role to tackle these effectively. In this sense, troubleshooting becomes an investigative process, akin to that of a detective. Design engineers must utilize critical thinking and problem-solving skills to analyze issues and arrive at viable solutions.

Before we wrap up, here's a final thought on engineering philosophy. As design engineers, we must remember: "The best is the enemy of the good." In other words, we need to know when to stop refining a design. While improvements can go on indefinitely, time and budget are finite. Striking the right balance is key to successful engineering.

2.2 Engineering Excellence: Evaluating Costs and Efficiency in Power Converter Design

Engineering fundamentally revolves around cost, making it a key consideration in the design and production of power converters and other products. The total cost of a power converter extends far beyond the immediate expense of the bill of materials (BOM). While component costs are an essential factor, numerous additional aspects contribute to the overall cost and must be thoroughly evaluated to understand the true financial footprint of the product.

Developing a power converter begins with research and development (R&D), which requires substantial investment in simulation, design, and prototyping. These expenses are not one-time but must be distributed over the product's expected lifecycle, ultimately influencing the cost per unit. Beyond the R&D phase, manufacturing overhead becomes a significant contributor. This encompasses labor, equipment upkeep, factory costs, and stringent quality control measures. Engineers must focus on optimizing designs for manufacturability to keep these expenses manageable. Strategies such as designing for automated assembly and reducing component complexity not only streamline production but also minimize the risk of defects, which can disrupt manufacturing efficiency and inflate costs.

Diagnostics and troubleshooting, both during development and after deployment, introduce additional costs. Complex diagnostic requirements can extend the time needed for testing and fault resolution, increasing labor expenses. In extreme cases, unresolved issues may lead to field failures, necessitating warranty repairs or replacements, both of which heavily impact the lifecycle cost. The risk of obsolescence also looms large over modern electronic designs. With the rapid pace of innovation in areas like semiconductors, components can quickly become obsolete. Engineers can mitigate this by adopting modular design practices, ensuring that individual parts can be replaced or upgraded without requiring a complete system overhaul. This approach extends the product lifecycle and shields against the high costs associated with redesigns. Warranty and after-sales support are further cost centers that cannot be overlooked. Products that fail within their warranty period typically require repair or replacement at the manufacturer's expense. As such, reliability becomes a critical design parameter. High-reliability designs lower the frequency of field returns and warranty claims, substantially reducing the long-term cost burden. Investments in rigorous testing, robust components, and well-established design practices pay dividends by preventing premature failures and enhancing the overall value proposition.

The relationship between cost and efficiency in power converters introduces another layer of complexity. This relationship, as depicted in Fig. 2.1, often follows a U-shaped curve. At one end of the spectrum, low efficiency increases costs due to the need for larger heat sinks, more elaborate cooling systems, and components capable of withstanding significant thermal stress. On the opposite end, attempting to approach near-100% efficiency leads to prohibitively high costs, as it requires advanced materials, hyper-optimized designs, and cutting-edge technologies. The

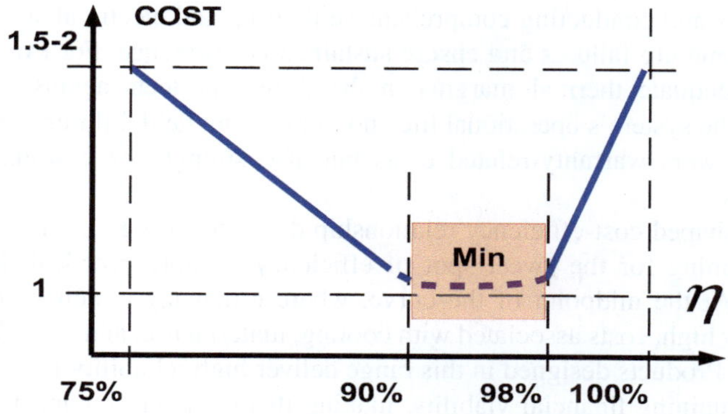

Fig. 2.1 The normalized cost of a converter versus power conversion efficiency

optimal efficiency range for most high-power converters lies between 90% and 98%, where the balance between performance and cost is most favorable. Efficiency should not be treated as an ultimate goal but rather as a variable to be optimized alongside cost. For most applications, the pursuit of extreme efficiency is only justified if the additional expense aligns with critical operational requirements, such as in aerospace or other high-stakes industries.

Switched-Mode Power Supplies (SMPS) are the cornerstone of achieving high efficiencies in energy conversion, particularly in the desired 90–98% range. Within SMPS, the choice between hard-switching and soft-switching significantly impacts both cost and efficiency. Hard-switching, while simpler and less expensive to implement, imposes considerable stress on semiconductors during transitions. This stress results in higher switching losses, substantial electromagnetic interference (EMI), and reduced component reliability. Mitigating these issues requires additional filtering and thermal management, which can offset the initial cost advantages. Over time, the cumulative impact of these drawbacks often translates to higher operational and maintenance costs.

Soft-switching, in contrast, offers a more sophisticated solution that minimizes the voltage and current overlap during switching transitions. This significantly reduces switching losses and EMI, leading to improved efficiency and simplified compliance with electromagnetic standards. Although soft-switching designs are more complex and require additional components, such as resonant circuits, the benefits outweigh the challenges in most high-power applications. Reduced thermal management needs also facilitate compact and lightweight designs, which are critical in industries like automotive and aviation, where space and weight constraints are paramount. By lowering material and maintenance costs, soft-switching enhances the overall cost-effectiveness of the system.

Reliability plays a pivotal role in cost reduction, reinforcing the value of soft-switching. While the initial investment in higher-quality components and more rigorous design processes may increase upfront costs, these measures drastically reduce lifecycle expenses. Reliability-focused practices, such as derating components and conducting comprehensive thermal and electrical stress analyses, prevent premature failures and ensure sustained performance. For example, incorporating adequate thermal margins in the design protects against overheating, extending the system's operational life and minimizing field failures. This approach not only lowers warranty-related costs but also strengthens customer trust and satisfaction.

The U-shaped cost-efficiency relationship depicted in Fig. 2.1 underscores this because aiming for the sweet spot of efficiency is more practical than chasing extremes. At the midpoint of the curve, where efficiency is neither too low nor excessively high, costs associated with cooling, material use, and manufacturing are optimized. Products designed in this range deliver high reliability and performance while maintaining financial viability, making them suitable for most commercial and industrial applications.

In conclusion, the choice between hard-switching and soft-switching technologies hinges on long-term cost considerations. While hard-switching may appeal due to its simplicity and lower initial costs, the drawbacks—such as increased EMI, higher thermal stress, and reduced reliability—often lead to higher overall expenses. Soft-switching, despite its complexity, emerges as the preferred solution in high-power converter applications. By delivering superior efficiency, reduced EMI, enhanced reliability, and lower maintenance costs, soft-switching aligns with the overarching engineering objective of cost-effective design. Ultimately, this balance of performance, reliability, and cost positions soft-switching as the optimal choice for achieving sustainable success in power conversion systems.

2.3 Parallel Connections in Power Converter Design: Cost Considerations and Challenges

One of the key engineering decisions that significantly influence the cost of power converter design is the use of parallel connections in power stages. This approach is often utilized when the design engineer must handle large currents and low voltages, especially in high-power applications such as automotive and industrial converters. When designing DC–DC converters, engineers often face another challenge: managing the ratio between input current and input voltage. This issue becomes even more pressing when delivering 2–4 kW of power at an input voltage of just 10–14 VDC, and the problem only gets worse as the power level rises or the voltage level drops.

Let's take a closer look at this with Fig. 2.2. This graph illustrates the relationship between the ratio of input voltage to input current (*V*/*A*) and the complexity of the design. Here's the takeaway: when this ratio is above one, the design is less challenging, and the higher it gets, the simpler the task becomes. On the other hand, if the ratio falls below one, things become increasingly difficult because circuit complexity grows, which makes it harder to meet design guidelines.

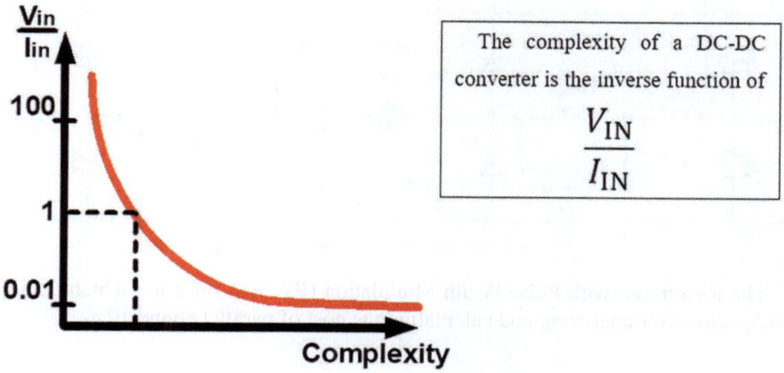

Fig. 2.2 The ratio of input voltage to input current (*V/A*) versus the complexity of the converter's design

To put it in perspective: designing a converter that operates at 10 VDC and 200 ADC is no small feat, but if we switch that to 200 VDC and 10 A, the task is significantly easier. To work around this complexity, engineers often turn to parallel power stages. However, paralleling stages introduces its own challenges, such as the need for precise current control. This can be achieved either through control loops or by relying on the intrinsic characteristics of the power stage.

Now, some engineers combine multiple power stages in parallel to achieve high-current and high-power ratings. This approach becomes essential when current ratings exceed 500 A or more. Below this threshold, though, the resulting power supply may not be cost-effective. Plus, this topology can struggle with reliability due to the increased number of switching devices and the challenge of sharing current evenly. Let's explore these points further.

Example Consider a step-down converter with the following specifications:

- Input voltage: 550–800 VDC
- Output voltage: 30 VDC
- Output current: 300 A
- Output power: 9 kW

In this scenario, the converter must handle very high currents to achieve the required output power. One method to address this problem is to parallel multiple power stages. Figure 2.3 shows a topology that provides inherent current sharing through the parameter characteristics of each power stage. This design will be discussed in detail in Chap. 6.

We will compare three different configurations:

1. A single power stage: 9 kW total power
2. Two paralleled power stages: Each stage is rated for 4.5 kW
3. Three paralleled power stages: Each stage is rated for 3 kW

Despite all configurations delivering the same total power of 9 kW, the cost and complexity of the designs vary significantly. Each implementation uses the same

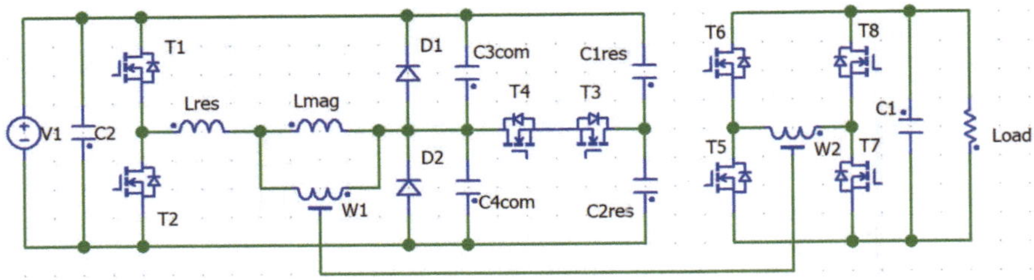

Fig. 2.3 The Resonance with Pulse Width Modulation (PWM) Converter, which is discussed in Section 6.5, is used for analyzing and calculating the cost of parallel connections

Table 2.1 Normalized cost comparisons of one power stage (PS) vs. two and three paralleled power stages in 2024

	One PS 9 kW	Two PS 4.5 kW × 2 = 9 kW	Three PS 3 kW × 3 = 9 kW
Overall cost	1	1.49	1.72
Transformer cost	0.38	0.38	0.33

topology, commutation frequency, and input/output voltage. The key difference lies in the number of paralleled stages.

Table 2.1 outlines the relative cost of each design. The design using a single power stage enjoys a significant cost reduction compared to those with two or three power stages. One major cost factor is the power transformer, which represents a substantial portion of the overall expense. Hence, the more complex designs with multiple stages and transformers result in higher costs.

It is also important to note that these calculations do not include the cost of labor, heat sinks, or physical enclosures. If we factor in these additional elements, the cost disparity between the configurations becomes even more pronounced. Additionally, this topology does not require controlling the current shared between power stages, as it inherently provides this functionality. In Chap. 6, we will explore this point in detail. Finally, adding current-sharing control would further increase cost and complexity. Therefore, we conclude that paralleling power stages often leads to higher costs and is not a cost-effective solution, especially in automotive applications where such designs are rare.

Note The calculations for Table 2.1 were performed in 2024. The authors performed similar calculations in 2016, with the results shown in Table 2.2. As we can see, the main proportions have not changed significantly; however, the cost contribution of the power transformer has increased. This indicates that, over the past 8 years, the cost of semiconductors has decreased relative to the rising cost of magnetic components. **This trend highlights the importance of design engineers focusing on the implementation of power transformers while aiming for a cost-effective product.**

Table 2.2 Normalized cost comparisons of one power stage vs. two and three paralleled power stages in 2016

	One PS 9 kW	Two PS 4.5 kW × 2 = 9 kW	Three PS 3 kW × 3 = 9 kW
Overall cost	1	1.53	1.83
Transformer cost	0.21	0.38	0.33

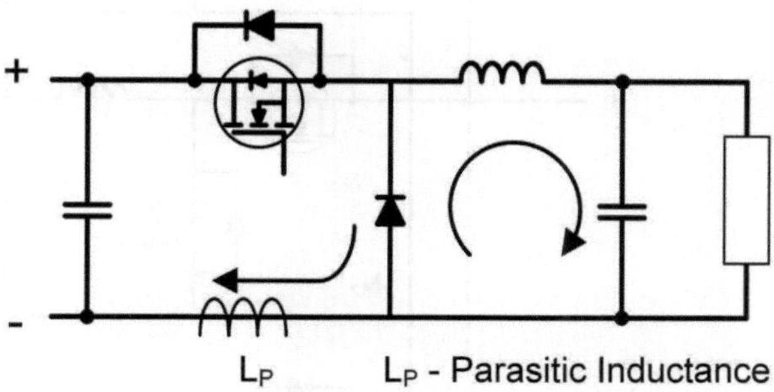

Fig. 2.4 A schematic of a buck converter where the stray inductance, L_P, creates parasitic transient current (PTC) in the main commutation loop

2.3.1 Paralleling and Control Challenges

In addition to the higher cost, paralleling power stages introduces additional challenges related to current control. The current sharing between stages must be carefully managed, either through active control loops or by leveraging the natural characteristics of the power stages themselves. When active control is required, the design becomes even more complex, necessitating more advanced algorithms and additional control hardware.

Furthermore, high-current designs with multiple stages also require robust current-sharing mechanisms to ensure an even distribution of load among the stages. If current-sharing is not achieved, some stages may become overloaded, leading to reduced reliability and even premature failure of components.

2.3.2 Isolated vs. Non-isolated Converters

Let us now consider a different aspect of the design decision—whether to use an **isolated** converter or a non-isolated one. This issue is equally important when designing converters with high current requirements, such as 48–12 VDC at 300 A, or 24–12 VDC at 100 A. In such designs, straightforward buck or boost converters often introduce significant parasitic circulation currents, leading to increased complexity in the EMI filter to meet EMI requirements. Figure 2.4 illustrates the basics of this problem. Some engineers attempt to address this issue by paralleling power stages and introducing phase-shifted commutation between stages to interleave the input and output currents. Figure 2.5 shows one option to mitigate parasitic

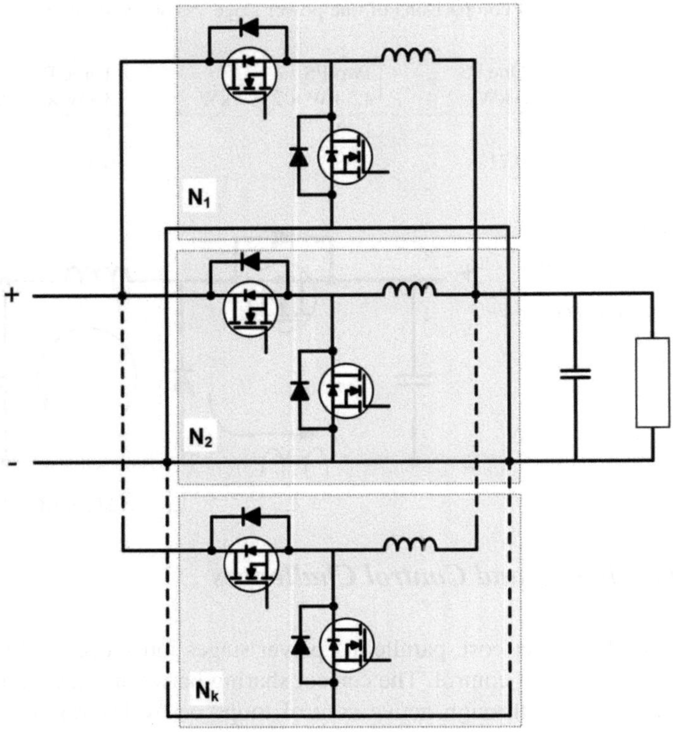

Fig. 2.5 Option 1—a parallel connection of smaller power stages that solves the parasitic transient current problem, but increases the cost and complexity of the converter

Fig. 2.6 Option 2—a configuration of an isolated DC-DC converter that mitigates parasitic transient current

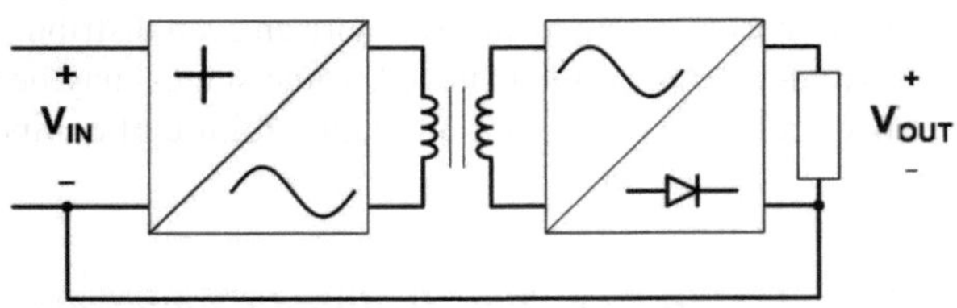

currents, but it comes at the cost of increased complexity, both in hardware and in the control strategy. Additionally, the design will still face challenges with current-sharing between stages, further complicating the overall system.

In contrast, isolated DC–DC topologies offer a more elegant solution. By incorporating isolation, these converters inherently reduce parasitic circulation currents and simplify the design of the EMI filter. Furthermore, isolated topologies do not suffer from the current-sharing issues that plague paralleled designs. As a result, an isolated DC–DC converter can achieve a cost-effective design while maintaining high reliability and lower EMI. Figure 2.6 illustrates this method, which offers the following advantages:

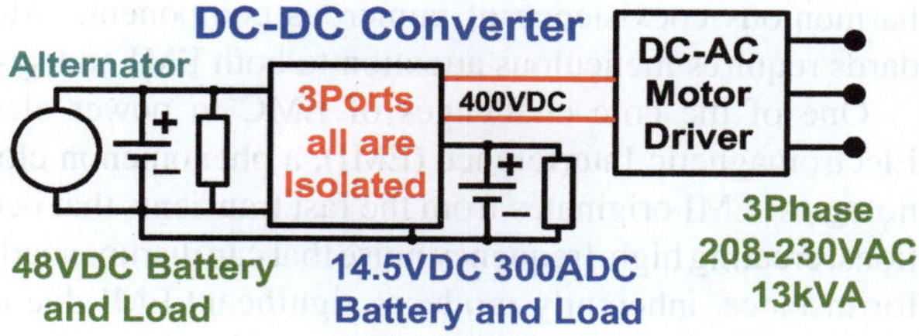

Fig. 2.7 An implementation of DC–DC converters for a vehicle electrification application that has three DC buses: a 14.5 VDC bus, a 48 VDC bus, and a 400 VDC bus

- Minimizes parasitic transient currents (PTC)
- Reduces the size and complexity of the EMI filter
- Simplifies control implementation
- Eliminates current sharing issues

These benefits collectively contribute to a reduction in overall cost.

2.3.3 Application in Automotive Industry

These concepts are widely applied in automotive applications today, especially in electric vehicles. Figure 2.7 shows a block diagram for a power unit designed for vehicle electrification. The DC–DC converter plays a key role, particularly in step-up mode, where the input current can reach as high as 350 ADC. Modern requirements for isolated DC–DC converters in automotive applications frequently demand power levels of 20–30 kW, with resulting currents of up to 600 ADC. In this implementation, the battery—though not galvanically isolated—works alongside an isolated DC–DC converter within the system, ensuring low EMI and cost efficiency. Further details on this type of converter will be explored later in the book.

2.4 The Symbiosis of EMC and Reliability in Power Electronics

Compatibility and reliability are two pillars of effective power converter design, particularly in high-performance applications like automotive and aviation. While this book does not center exclusively on Electromagnetic Compatibility (EMC), it is impossible to address isolated DC–DC converters without delving into this critical subject. Over the past 20 years, EMC requirements have tightened considerably, especially for automotive systems. Where compliance with CISPR25 Class 2 standards was once sufficient, modern vehicles now demand adherence to CISPR25 Class 4 standards or higher [1]. This evolution reflects the increasing complexity of electronic systems within vehicles, where seamless operation depends on the harmonious coexistence of numerous components. Meeting these stringent standards requires meticulous attention to both EMI and system reliability.

One of the core challenges of EMC in power electronics lies in controlling Electromagnetic Interference (EMI), a phenomenon closely tied to switching technologies. EMI originates from the fast transients that occur during switching transitions, creating high-frequency noise that can disrupt nearby systems. Hard-switching, for instance, inherently produces significant EMI due to the high voltage and current overlap during transitions. Conversely, soft-switching offers a clear advantage by minimizing this overlap, thereby reducing the transients responsible for EMI. This reduction not only simplifies the design and cost of EMI filters but also enhances the reliability of the entire system. By limiting EMI at the source, engineers can achieve compliance with regulatory standards while ensuring stable operation. In addition to switching methodologies, slew rate control plays a critical role in managing EMI. For power devices like MOSFETs and IGBTs, the rate of change of voltage over time (*dV*/*dt*) is a primary determinant of radiated EMI. High slew rates, particularly those exceeding 10,000 V/μs, can significantly amplify EMI, necessitating more robust filtering and shielding solutions. A more moderate slew rate, ideally below 5000 V/μs, strikes a balance between switching speed and EMI reduction. Controlling *dV*/*dt* not only simplifies EMI mitigation but also has a profound impact on reliability, as it reduces stress on both the switching components and connected systems.

The interplay between high *dV*/*dt* and reliability is well-illustrated by a historical example from the late 1980s. When IGBTs began replacing traditional transistors and thyristors in industrial settings such as mining, engineers observed an unexpected increase in motor winding failures. These failures were traced to the fast-switching characteristics of IGBTs, which produced transients at frequencies that resonated with the natural wavelength of motor wiring harnesses. This resonance turned the wiring into an unintentional transmission line, causing voltage spikes that could double the DC bus voltage [2, 3]. The resultant stress led to premature breakdowns in motor windings, highlighting the critical need for *dV*/*dt* control in high-reliability designs.

Such cases underscore the importance of addressing transient stresses in power systems. High *dV*/*dt* and fast transients not only stress switching components but also impact peripheral equipment like motors, transformers, and inductors. For isolated DC–DC converters, this translates into increased wear and tear on key components, thereby reducing the overall system reliability. Effective mitigation strategies, such as soft-switching and proper circuit layout, are essential for controlling transients and extending the operational life of these systems.

Reliability is particularly critical in industries like transportation and aviation, where power electronics are expected to operate for 20 years or more without failure. Aviation, in particular, imposes extreme reliability standards, as human lives depend on the performance of these systems. This has led to the adoption of rigorous design practices, such as component derating, which involves operating components well below their maximum rated limits. By reducing thermal, electrical, and mechanical stress, derating significantly enhances the durability and reliability of the system.

Soft-switching technology contributes substantially to reliability by reducing transient stresses during operation. Unlike hard-switching, which generates high losses and stresses at each transition, soft-switching ensures that voltage and/or current are near zero during switching events. This dramatically reduces both thermal and electrical stress on components like MOSFETs, IGBTs, and diodes, extending their lifespan and improving system robustness. Additionally, the reduced heat generation associated with soft-switching simplifies thermal management, further enhancing reliability.

The mutual relationship between EMI and reliability cannot be overstated. Excessive EMI not only complicates compliance with regulatory standards but also jeopardizes the stability of the power converter and adjacent systems. High EMI levels can induce intermittent faults, degrade performance, and even lead to catastrophic failures over time. By addressing EMI through soft-switching and controlled slew rates, engineers create systems that are both more reliable and easier to certify for demanding applications. Furthermore, reduced EMI means fewer resources need to be allocated to filtering and shielding, lowering costs while maintaining performance.

This interplay is particularly relevant in the context of automotive and aviation systems, where the demands for reliability and compatibility converge. In automotive systems, for example, power converters must operate seamlessly alongside communication networks, infotainment systems, and advanced driver-assistance systems (ADAS). Achieving this level of integration requires designs that not only meet stringent EMC standards but also offer long-term reliability under harsh operating conditions. Similarly, in aviation, the consequences of EMI-induced failures are far more severe, making the case for robust, reliable designs even stronger.

In conclusion, focusing on EMC and reliability in power converter design yields multiple benefits. Reduced EMI leads to simpler and more cost-effective filtering solutions, while improved reliability lowers maintenance costs and extends product lifetimes. The adoption of soft-switching technology and stringent dV/dt control are essential strategies for achieving these goals. By prioritizing compatibility and reliability, engineers can create power converters that meet the highest standards of performance, safety, and cost-effectiveness, addressing the complex needs of modern automotive and aviation applications.

Note The authors strongly believe that the complexity of the topology and the surrounding system, including auxiliary power supplies, is significantly affected by the effort to minimize EMI (Electromagnetic Interference) as much as possible. Designing a system with low EMI from the beginning makes it much easier to finalize the product. In contrast, starting with a prototype that exhibits high EMI often leads to a prolonged and costly process of addressing EMI issues through filtering and other adjustments. This effort not only consumes additional time but also increases the overall cost. In other words, it is far better to address the root causes of EMI during the design phase than to deal with its consequences later.

2.5 The Art of Compromise: Managing Design Trade-Offs in Engineering

Trade-offs in design objectives are an inherent part of any engineering project, requiring careful deliberation to balance multiple competing priorities. Each design decision represents a compromise, where improving one aspect often comes at the expense of another. This balancing act becomes especially critical in the development of complex systems such as isolation DC–DC converters and other power electronics, where factors like performance, cost, reliability, and time-to-market are interdependent and demand constant evaluation. The ultimate goal is to find a practical solution that meets the design objectives while staying within the constraints of the project.

The engineering process begins with a thorough identification of potential weaknesses in the design. These weaknesses can manifest as vulnerabilities that compromise performance or reliability. Examples include components that fail under high stress, inefficiencies in power conversion processes, excessive heat generation that complicates thermal management, or the inability to comply with electromagnetic interference (EMI) regulations. By pinpointing these weak points early, engineers can focus on mitigating risks, often through redesigns, component selection, or optimization of operating conditions. Ignoring such weaknesses during the initial phases can lead to significant problems later, including costly revisions, production delays, or even product recalls.

Equally important is the evaluation of a design's strengths, which highlights the aspects that excel in meeting key objectives. These strengths often reflect the design's ability to deliver on critical criteria, such as efficiency, compactness, manufacturability, or cost-effectiveness. For instance, a specific converter topology might provide outstanding efficiency under typical operating conditions but could involve higher costs due to more complex components or manufacturing requirements. Understanding these strengths allows engineers to make informed trade-offs, emphasizing areas where the design adds the most value.

The decision-making process in engineering is essentially a balancing act that weighs these strengths and weaknesses. Engineers must make technical compromises that align with the project's overall priorities. This often involves choosing between competing factors, such as efficiency versus cost, size versus power capability, or reliability versus speed of development. These trade-offs are central to the engineering discipline, as they define how well a design achieves its intended purpose while adhering to project constraints.

Amid these considerations, cost emerges as a critical and often dominant factor. While performance and reliability remain vital, the ability to manage costs effectively can determine a product's commercial viability. Cost considerations span the entire lifecycle of a product, from research and development (R&D) to production and after-sales support. Managing these costs requires a strategic approach to resource allocation, balancing immediate expenses with long-term benefits.

During the R&D phase, costs are primarily driven by the need for innovation and validation. Skilled personnel, including engineers and technicians, must dedicate significant time to developing, prototyping, and testing the product. Materials used for prototypes and simulations also contribute to R&D expenses. Additionally, access to specialized facilities, such as testing laboratories and manufacturing resources, represents a substantial investment. Validation testing, which includes stress tests, thermal performance evaluations, EMI compliance checks, and lifecycle analysis, is crucial to ensuring the product meets all specifications. These activities, while costly, are necessary to minimize risks during production and deployment.

When the product enters production, cost considerations shift to manufacturing and distribution. The bill of materials (BOM), which includes all components necessary to build the product, becomes a key driver of expenses. Decisions about component selection—such as opting for high-quality semiconductors that reduce switching losses but come at a premium—directly impact both performance and cost. Manufacturing efficiency, labor costs, and quality control processes further influence the overall cost of production. Distribution logistics, including shipping and handling, add another layer of complexity, especially when products must reach global markets.

After-sales costs, though often underestimated during the design phase, significantly affect a product's long-term success. Installation expenses can be substantial for complex power electronics, particularly when specialized expertise is required. Simplifying installation through standardized interfaces or modular designs can help reduce these costs. Design revisions are another potential expense, as unforeseen issues may necessitate modifications after the product has been deployed. Poorly executed designs often lead to higher rates of warranty claims, product recalls, or retrofitting, which can erode profitability. Additionally, products with higher failure rates increase costs associated with warranties, service, and repair, making reliability a critical focus during the initial design process.

One of the most challenging trade-offs in power electronics design is between cost and reliability. High reliability often demands the use of robust components, careful derating, and extensive testing. These measures drive up initial costs but significantly reduce maintenance and repair expenses over the product's lifecycle. For instance, isolation DC–DC converters designed for automotive applications must often maintain reliability for up to two decades. Any failure in such converters could compromise the entire vehicle's functionality. To enhance reliability, engineers employ techniques like soft switching to reduce stress on components, thereby extending their operational lifespan. Although this increases upfront costs, it minimizes total ownership costs by reducing downtime and repair needs.

Design decisions have a far-reaching impact, extending beyond the immediate product development phase. For example, choosing lower-cost components might seem advantageous initially, but it could lead to increased assembly complexity, higher thermal management requirements, or reduced reliability. Each of these outcomes adds costs elsewhere in the product lifecycle, whether during production, customer support, or warranty fulfillment. Thus, the trade-off is not simply between performance and cost—it encompasses a broader evaluation of short-term savings versus long-term implications. A well-balanced design optimizes the total lifecycle cost, ensuring that the product remains competitive in terms of both functionality and affordability.

Ultimately, engineering is a continuous exercise in balancing trade-offs between competing objectives. While performance and reliability are essential, cost considerations often drive final design decisions. Engineers must carefully evaluate factors spanning R&D, production, and after-sales support to create products that deliver reliable performance while remaining economically viable. The key to success lies in achieving a harmonious balance—a design that meets technical requirements, upholds reliability, and remains cost-effective over its entire lifecycle.

2.6 Conclusion

This chapter presented our Engineering Philosophy and how it applies to product design and development. We provided strategies to guide the design engineer toward a successful project and viable commercial product by prioritizing sound engineer judgment, which begins with understanding the mission priorities of the project. The design engineer must always consider the system as a whole when executing their designs.

2.7 Supplemental

The content of this chapter represents the original, creative work of the authors. We utilized OpenAI's ChatGPT to check grammar and spelling and to improve the clarity of certain portions of the text. Portions of the text and images are from tutorials that we presented at the *2017 IEEE Energy Conversion Congress and Exposition (ECCE 2017)*, Cincinnati, OH, USA, 2017 and *2022 IEEE Energy Conversion Congress and Exposition (ECCE 2022)*, Detroit, MI, USA, 2022.

Additionally, some of the images and data contained in this chapter were provided by Havis, Inc. with support from Alec Cook.

References

1. CISPR, CISPR 25:2021 (or EN 55025:2021), Vehicles, boats and internal combustion engines—Radio disturbance characteristics—Limits and methods of measurement for the protection of on-board receivers, 4th edn., 16 Dec 2021

2. M.J. Scott et al., Reflected wave phenomenon in motor drive systems using wide bandgap devices, in *2014 IEEE Workshop on Wide Bandgap Power Devices and Applications, Knoxville, TN, USA*, (2014), pp. 164–168. https://doi.org/10.1109/WiPDA.2014.6964647
3. M.J. Scott et al., Design considerations for wide bandgap based motor drive systems, in *2014 IEEE International Electric Vehicle Conference (IEVC), Florence, Italy*, (2014), pp. 1–6. https://doi.org/10.1109/IEVC.2014.7056214

Chapter 3
From Theory to Practice: Meeting Design Specifications for Isolated DC–DC Converters

The primary responsibility of power converters is to seamlessly transfer energy from a source, or multiple sources, to a load or multiple loads. In this chapter, we provide a high-level overview of several key design considerations for this technology. We begin by examining the trade-offs between isolated and non-isolated DC–DC converters. Since this book focuses exclusively on power electronics, with a high-voltage side operating at several hundred volts and a low-voltage side carrying several hundred amps, isolated DC–DC converters are the most economical choice in terms of cost, reliability, and compatibility.

Next, we explore the operation of semiconductor devices. While design engineers have control over how these components are utilized, we analyze the trade-offs between hard-switching and soft-switching techniques in power conversion hardware. Although soft-switching technology is more complex to implement than hard-switching, it is generally more efficient and reliable. Furthermore, when considering the overall system—along with manufacturing and warranty costs from a business perspective—soft-switching often proves to be the most cost-effective approach.

There are two primary forms of soft-switching technology: zero-voltage switching (ZVS) and zero-current switching (ZCS). We examine both approaches and outline the advantages and disadvantages of each. Additionally, we show the disadvantage of a passive snubber circuit. For this purpose, we introduce a lossless snubber, detailing its operation and key design considerations.

We conclude the chapter by discussing the three operating modes of a power converter: constant voltage, constant current, and constant power. A special case of a constant power source, known as discrete energy transfer, is highlighted due to its widespread application in resonant converters.

Ultimately, design engineers must consider the loads their equipment will serve and the specific requirements of the hardware's application. This ensures the development of equipment that meets the demands of cost, reliability, and compatibility.

A. Isurin, M. J. Scott, *Practical Design Considerations for Isolated DC-DC Converters*, https://doi.org/10.1007/978-3-031-94427-7_3

3.1 Motivation for Using Isolated DC–DC Converters

Electric power converters, which perform energy conversion processes, are central to a vast array of modern technologies. Their widespread application spans everything from industrial equipment to renewable energy systems, electric vehicles, and consumer electronics. Among the various types of power converters, isolation DC–DC converters hold a unique position due to their ability to not only transfer energy but also provide galvanic isolation. This book explores the behavior, characteristics, and practical considerations of isolation DC–DC converters.

Rather than delving into the broader realm of power conversion, which has been extensively studied and documented, this work narrows its focus to the specific challenges and design strategies associated with isolation DC–DC converters. While general principles of power conversion are relevant to understanding their operation, the intent here is to address the nuances that differentiate isolated converters from their non-isolated counterparts.

The primary purpose of an isolation DC–DC converter is to transfer energy from a source to a load efficiently while maintaining complete electrical separation between the two. Galvanic isolation plays a critical role in numerous applications. It prevents the direct flow of current between input and output circuits, enhancing safety, protecting sensitive electronic systems, and eliminating ground loops that can introduce noise or compromise signal integrity. Furthermore, isolation allows for flexibility in system design, enabling converters to handle widely differing input and output voltage levels without compromising performance or safety.

In technical terms, an isolated DC–DC converter performs its function using a two-stage process. Figure 3.1 shows the block diagram for this process. The first stage is a DC–AC conversion, in which the input DC voltage is converted into a high-frequency alternating current (AC) waveform. The second stage is an AC–DC conversion, where the transformed AC waveform is rectified back into DC voltage at the output. A power transformer situated between these two stages not only facilitates voltage transformation but also provides crucial galvanic isolation.

Both stages operate in switch mode, using high-speed electronic switches to achieve efficient energy conversion. When ideal conditions are assumed—such as zero switching times, zero dead times, and no parasitic effects—a theoretical 100% efficiency is achievable. Under these circumstances, energy transfer occurs seamlessly, and the input and output currents flow continuously, without interruptions or

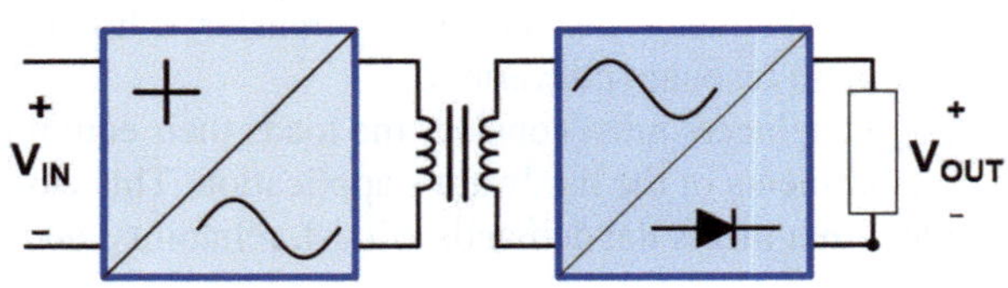

Fig. 3.1 The block diagram of an isolated DC–DC converter with two stages: a high-frequency inverter and a high-frequency rectifier

losses. This idealized scenario represents the ultimate goal for engineers designing isolation DC–DC converters.

In practice, however, deviations from this ideal are unavoidable. Real-world designs must contend with non-zero switching times, dead times, and various forms of losses, including conduction and switching losses. These factors reduce the effective duty cycle and overall efficiency of the converter. Despite these challenges, modern designs can achieve duty cycles exceeding 90%, which is often sufficient for many demanding applications.

One of the most significant advantages of isolated DC–DC converters is their ability to achieve high voltage conversion ratios. Unlike non-isolated converters, such as Buck, Boost, or Buck-Boost topologies, which are typically constrained by practical voltage conversion limits, isolated converters can accommodate voltage ratios exceeding three or more. This capability is particularly beneficial in applications requiring significant voltage step-up or step-down, such as renewable energy systems, electric vehicle chargers, and high-voltage industrial equipment.

The decision to use an isolated versus a non-isolated design often sparks debate among engineers, especially in high-power applications where galvanic isolation is not strictly required. The inclusion of isolation introduces additional components and complexity, which can increase cost and reduce overall efficiency. However, in applications where safety, reliability, or regulatory compliance is a concern, the benefits of isolation frequently outweigh these drawbacks.

Understanding the behavior of an isolation DC–DC converter requires a detailed examination of its losses. Conduction losses, caused by the inherent resistance of the conduction paths within switches and passive components, are among the most significant contributors to inefficiency. These losses result in heat generation, necessitating robust thermal management strategies to maintain reliability and performance. Engineers must carefully select components, optimize circuit layouts, and employ advanced materials to minimize these losses.

Switching losses play a critical role in real-world designs. These losses occur during the transitions between the on and off states of the electronic switches. Techniques such as zero-voltage switching (ZVS) and zero-current switching (ZCS) are widely used to mitigate these losses, particularly in high-frequency converters.

Achieving the ideal performance of an isolation DC–DC converter involves meeting several aspirational criteria. These include:

1. **100% efficiency**: The converter should ideally transfer all input energy to the output without any losses. While this is impossible in practice, maximizing efficiency remains a primary design objective.
2. **Perfect galvanic isolation**: Ensuring complete electrical separation between input and output circuits enhances safety and operational stability.
3. **Bi-directional power flow**: Many modern applications, such as energy storage systems, require converters to support power flow in both directions, enabling efficient energy management.

4. **Fast response to load variations**: An ideal converter would instantly adapt to changes in load demand, maintaining stable output conditions without delays or fluctuations.
5. **Wide input voltage range**: Flexibility in handling varying input voltages while delivering consistent output performance is crucial for many applications.
6. **High power density**: Compact designs that optimize power delivery while minimizing size and weight are highly desirable, particularly in space-constrained environments.
7. **Minimal electromagnetic interference (EMI)**: Reducing EMI ensures compliance with regulatory standards and prevents interference with other electronic systems.
8. **Low output ripple**: Clean output voltage and current are essential for powering sensitive electronic components and ensuring system reliability.

Each of these characteristics represents a design goal that engineers strive to balance in practical implementations. However, compromises are often necessary. For example, achieving high power density might require increasing switching frequencies, which can introduce new challenges in EMI and thermal management. Similarly, optimizing for low cost might limit the ability to use advanced materials or components, affecting performance in other areas.

In practical designs, the transformer plays a central role in determining the overall performance of the isolation DC–DC converter. Its size, core material, and winding configuration directly influence efficiency, voltage transformation, and isolation quality. Advances in transformer technology, such as the use of high-frequency magnetic materials and special winding structures, have significantly improved the performance of modern converters.

Thermal management is another critical aspect of design. The heat generated by conduction and switching losses must be effectively dissipated to prevent overheating and ensure long-term reliability. Solutions range from passive methods, such as heatsinks and natural convection, to active cooling systems, such as forced air or liquid cooling. Selecting the appropriate cooling strategy depends on the application's specific requirements and constraints.

Finally, the debate over isolation versus non-isolation reflects broader considerations about system design priorities. While non-isolated converters offer simplicity and cost advantages, isolated converters provide unmatched flexibility, safety, and performance in many scenarios. For applications where safety and compliance with strict standards are mandatory, isolation becomes an indispensable feature.

In conclusion, the isolation DC–DC converter is a versatile and essential component in modern power electronics. Its ability to efficiently transfer energy while providing galvanic isolation makes it indispensable in applications ranging from renewable energy systems to aerospace and industrial automation. While achieving the ideal performance remains an aspirational goal, the principles and strategies outlined here provide a foundation for advancing the design and optimization of these critical devices. Engineers must navigate the complex interplay of efficiency, reliability, and cost to create solutions that meet the diverse needs of contemporary applications.

3.2 An Overview of Semiconductor Switching Operation

In discussing the implementation of DC–DC converters, particularly through electronic switch mode power conversion, it is essential to delve into the switches themselves and their operational mechanics. The core of these switches is embedded in semiconductor technology. In Chap. 4, we will explore the various types of semiconductors currently available, their distinct characteristics, and the considerations necessary for their effective drive and control.

Switching Behavior in Power Devices
The switching behavior of power devices in a switch-mode power supply (SMPS) can be classified as either hard-switching or soft-switching. Given the extensive literature on electronic switches, it may seem unnecessary to reiterate basic concepts; however, we will focus on the practical implementation of these switches, particularly as they relate to isolated DC–DC converters.

3.2.1 Hard-Switching

Hard-switching is a common technique used in DC–DC converters due to its straightforward implementation and relatively simple control requirements. However, it comes with a variety of challenges that can impact overall system performance, reliability, and cost. Understanding these factors is crucial for designing effective power conversion systems.

3.2.1.1 Key Challenges and Cost Implications

1. **Increased component stress**: During hard-switching, semiconductors experience substantial stress during the transition between on and off states. This stress can lead to increased thermal management needs, as higher temperature operation can degrade the materials over time. As a result, engineers may opt for higher-rated components that can withstand these conditions, leading to increased costs.
2. **Voltage and current spikes**: The high levels of dV/dt (rate of voltage change) and dI/dt (rate of current change) associated with hard-switching can produce voltage spikes and ringing, which may necessitate additional components such as filters and snubbers to manage EMI. These additional components not only

contribute to the overall material cost but also increase the complexity of the design and assembly process.

3. **Electromagnetic interference (EMI)**: hard-switching generates substantial EMI due to the high rates of *dV/dt* and *dI/dt* during switching transitions. This stress not only affects the semiconductors themselves but also places a considerable strain on passive components, particularly capacitors, which must operate in conjunction with the semiconductor devices. The consequence of this stress is a reduction in the reliability and lifetime of the converter. Additionally, regulatory compliance with EMI standards can necessitate more rigorous testing and validation processes, leading to higher developmental costs.
4. **Efficiency trade-offs**: Hard-switching converters typically operate with lower efficiency than their soft-switching counterparts, particularly at high frequencies. Lower efficiency often means more power dissipated as heat, which requires additional thermal management solutions (such as heat sinks or active cooling) that can add to both the initial cost and long-term operational expenses.
5. **Size and weight considerations**: The components required to manage the consequences of hard-switching can lead to larger PCB layouts and heavier designs. For applications where size and weight are critical factors, such as in aerospace or portable electronics, these increased dimensions can significantly impact the feasibility and cost-effectiveness of the product.

3.2.1.2 Conclusion

While hard-switching techniques offer simplicity in design and control, the associated challenges can significantly impact the overall cost and reliability of DC–DC converters. Designers must carefully weigh these factors against the specific requirements of their application. A thorough understanding of the implications of hard-switching enables engineers to make informed decisions regarding component selection, circuit design, and system architecture, ultimately guiding them toward cost-effective and reliable power conversion solutions.

To mitigate the adverse effects of hard-switching, design engineers often incorporate snubber circuits, which add complexity to the implementation process. Notably, these circuits are located outside the transistor package, leaving the semiconductor die exposed to stress. This makes it critical to address these issues with close attention.

3.2.2 Soft-Switching

Soft-switching, by contrast, is inherently more complex but significantly reduces—or even eliminates—switching losses and mitigates EMI by decreasing the slew rate (*dV/dt*). A key advantage of soft-switching is its ability to eliminate the need for complex snubber circuits, as the switching process is inherently managed.

Additionally, soft-switching facilitates the integration of the control stage with the power stage, reducing the impact of power processing on control signals. As a result, designs utilizing soft-switch technology often achieve lower overall costs compared to those employing hard-switch technology. Thus, adopting soft-switch technology not only enhances performance and reliability but also leads to cost-effective designs.

In summary, while hard-switching offers simplicity, its drawbacks necessitate careful consideration in the design process. Soft-switching, despite its complexity, provides significant benefits in reducing stress on components and associated costs, making it an attractive option for efficient and reliable converter designs.

Soft-switching techniques in power electronics are primarily categorized into two forms: Zero Voltage Switching (ZVS) and Zero Current Switching (ZCS). While both methods share similar advantages, each presents unique challenges that warrant further discussion.

3.2.2.1 Zero Voltage Switching (ZVS)

ZVS occurs when the switching device turns off or on while the voltage across it is zero. This characteristic allows for minimal stress on the semiconductor, thereby enhancing its reliability.

Figure 3.2 illustrates the simplified behavior of ZVS. This method is commonly implemented in various converter topologies, including phase-shifted converters and Dual Active Bridge (DAB) converters. During ZVS operation for a phase-leg topology, consider a scenario where the upper switch turns off and the lower switch turns on. Under optimal conditions, both switches can achieve ZVS:

- When the upper switch turns off, the voltage across capacitor C_2 cannot change instantaneously due to its capacitance. If C_2 is sufficiently large, it will enable the upper switch to turn off under ZVS conditions.

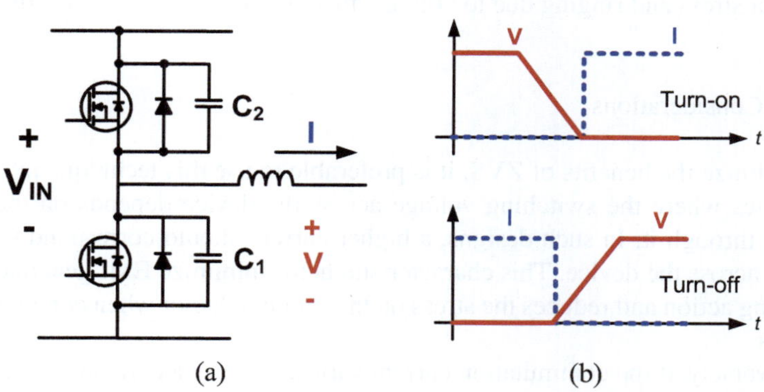

Fig. 3.2 The simplified behavior of zero voltage switching (ZVS) with (a) the schematic of a half-bridge configuration and (b) its corresponding switching waveforms for the turn-on and turn-off conditions

- For the lower switch, the current flowing through the inductive element discharges the output capacitance, C_1, while the switch is off. Once the voltage across the device goes to zero, the anti-parallel diode conducts, allowing the lower switch to turn on under ZVS conditions.

It's worth mentioning that the inductor may be composed of the leakage inductance from the transformer, a specially designed inductor, or both.

Disadvantages of ZVS

While ZVS offers significant advantages, it has some limitations:

1. **Dependency on commutation current**: The primary disadvantage of ZVS lies in the dependence of its switching time on the value of the commutation current. This means that the *dV/dt* across the power device becomes variable; a higher current results in a higher *dV/dt*, whereas a lower current leads to a lower *dV/dt*. For instance, using established relationships, we can calculate the switching time as the switch's voltage multiplied by the sum of the two parallel capacitances, divided by the current in the inductor. For example, if $V_{IN} = 400$V and $C_1 + C_2 = 0.02\,\mu$F, a current of 50 A achieves ZVS in 0.16 μs, whereas a current of 5 A accomplishes this in 1.6 μs. As a result, at higher currents, the switch enters hard switch mode. This behavior of switch commutation significantly influences the implementation of the converter when it needs to provide a wide range of operating conditions.
2. **Capacitance and additional components**: In practical applications, the presence of bond wire inductance can further complicate ZVS operation. If the commutation current is low but the body capacitance is adequate, ZVS can be effectively maintained. However, if the commutation current is high and the body capacitance is insufficiently small, additional commutation capacitors may need to be connected externally to the transistor. This setup can introduce additional stress and ringing due to voltage fluctuations on the transistor die.

Design Considerations

To maximize the benefits of ZVS, it is preferable to use this technique in converter topologies where the switching voltage across the device depends on the current flowing through it. In such designs, a higher current should correspond to a lower voltage across the device. This characteristic helps minimize EMI generated by the switching action and reduces the stress on the semiconductor when conducting high currents.

Conversely, if the commutation current varies, creating a corresponding variable commutation voltage would be beneficial. Alternatively, the changes in commutation current should remain small to maintain effective ZVS operation.

Conclusion

In summary, while ZVS provides enhanced reliability and lower EMI, its effectiveness heavily depends on commutation current and capacitance considerations. Understanding these relationships is key for designers seeking to implement ZVS successfully in DC–DC converters.

3.2.2.2 Zero Current Switching (ZCS)

ZCS is a technique used in power electronics to enable the switching of devices when the current through them is zero. This approach significantly reduces the switching losses associated with the transition of power devices because the device turns on without the presence of current, thereby minimizing stress on the components. As demonstrated in Fig. 3.3, this behavior occurs in circuits involving inductive elements where the current cannot change instantaneously.

Principles of ZCS Operation

In ZCS operation, when the switching device is activated, the current flowing through the inductors remains stable at zero. Here's how the operation unfolds:

1. **Inductor behavior**: The inductors in the circuit play a critical role in maintaining current flow. Because the current through the inductors can't drop instantaneously, the device remains off until the current naturally decays to zero.
2. **Turning on the device**: Once the current reaches zero, the device can safely turn on without experiencing voltage spikes. This condition helps avoid the issues associated with hard-switching, such as EMI and increased thermal stress.

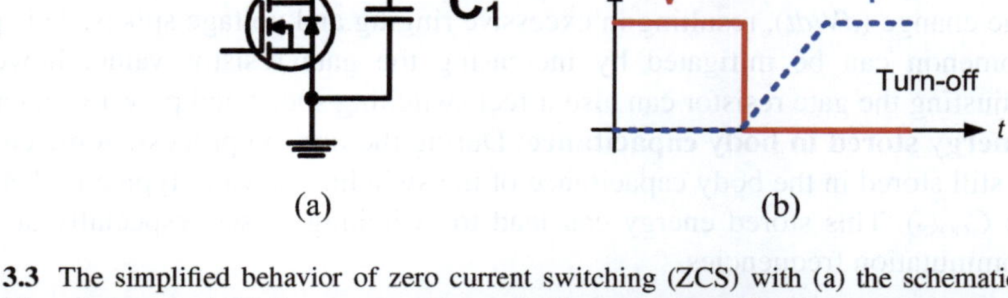

Fig. 3.3 The simplified behavior of zero current switching (ZCS) with (a) the schematic of a switching devices, and (b) its corresponding switching waveforms for the turn-on and turn-off conditions

3. **Application of forced commutation**: The principles of ZCS have historical roots, originating from techniques used during the 1960s and 1970s in thyristor circuits and other forced commutation systems to achieve controlled turn-off of devices.

Applications of ZCS

ZCS is particularly useful in various applications, including:

- **Resonant converters**: ZCS is frequently employed in resonant converters to take advantage of the benefits of soft-switching while controlling the energy transfer efficiently.
- **Inductive heating**: The technique finds applications in systems requiring controlled power delivery that avoids sudden changes in current.
- **High-power inverters**: ZCS can improve efficiency and reliability in inverters for renewable energy systems, electric vehicles, and motor drives.

Advantages of ZCS

- **Reduced switching losses**: The absence of commutation current during the switching event greatly reduces the energy lost as heat in the power devices, contributing to higher overall efficiency.
- **Lower EMI generation**: Similar to ZVS, ZCS minimizes voltage spikes and ringing during the switching process, leading to lower EMI and potentially less stringent electromagnetic compliance issues.
- **Enhanced reliability**: By reducing switching stress on devices, ZCS contributes to longer component lifetimes and improved reliability in power electronics systems.

Challenges and Limitations of ZCS

Despite its advantages, ZCS also presents several challenges:

1. **High *dV/dt* and ringing**: One significant disadvantage of ZCS is related to switching time. If the switching time is too short, it can lead to high rates of voltage change (*dV/dt*), resulting in excessive ringing and voltage spikes. This phenomenon can be mitigated by increasing the gate resistor value; however, adjusting the gate resistor can also affect switching speed and power efficiency.
2. **Energy stored in body capacitance**: During the turn-on process, some energy is still stored in the body capacitance of the switching device (typically denoted as C_{OOSS}). This stored energy can lead to switching losses, especially at high commutation frequencies.

3. **Commutation frequency limitations**: If the commutation frequency is too high, the cumulative losses can become unacceptable, making it challenging to optimize performance for high-frequency applications. Such limitations can restrict the practical implementation of pure ZCS configurations.
4. **Transition to ZVS**: When small commutation currents are utilized, ZCS can transition into ZVS mode. While this provides some flexibility, it may complicate the design and performance characteristics of the power converter.

Conclusion

ZCS is a valuable technique that offers distinct advantages in reducing switching losses and EMI. However, it also comes with its own set of challenges, particularly concerning ringing, switching time, and practical frequency limits. Understanding the trade-offs and inherent constraints of ZCS is essential for engineers designing efficient and reliable power converters. As we continue to explore power electronics techniques, careful consideration of ZCS and its implementation will be critical in optimizing performance across various applications.

3.2.3 Lossless Snubbers and Soft-Switching: Solutions for High-Performance DC–DC Converters

Let's switch gears and discuss lossless snubbers, essential components in power electronics that manage switching events in DC–DC converters. These devices play a critical role in improving efficiency, reducing EMI (Electromagnetic Interference), and protecting components from voltage spikes. Lossless snubbers come in a variety of implementations, both passive and active. A common feature among them is the incorporation of diodes and/or transistors that operate in ZCS mode.

3.2.3.1 Operational Dynamics

To understand the behavior of lossless snubbers, consider the case of a **passive snubber** as illustrated in Fig. 3.4. In practical applications, several factors influence the circuit's performance:

1. **Parasitic inductance**: Diodes used in lossless snubbers inherently exhibit parasitic inductance (see L_P in Fig. 3.4), arising from bond wires, lead connections, and stray inductances on printed circuit boards (PCBs). These inductances significantly affect the switching behavior of the circuit.
2. **Current rise time**: When the diode begins conducting, the **rise time of the current** plays a pivotal role in determining performance. At higher commutation currents, the current rise time through the diode increases. Conversely, the

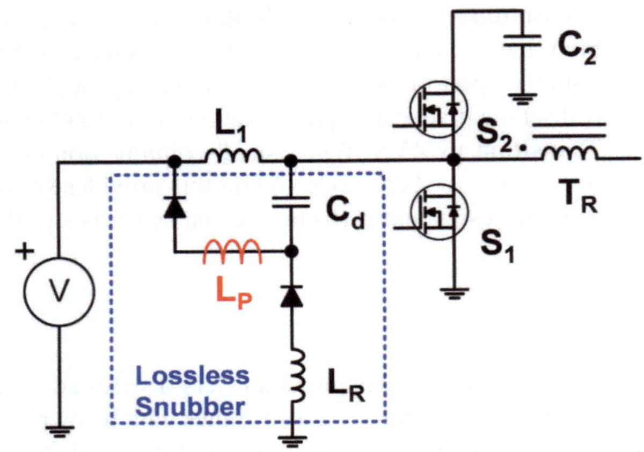

Fig. 3.4 An illustration of how the parasitic inductance (L_P) from bond wires, lead connections, and stray inductances on the PCB limits the performance of a lossless snubber

transistor's switching time may decrease with rising current, potentially causing a mismatch where the transistor switches off before the diode fully conducts. This condition leads to **hard-switching**, undermining the advantages of the lossless snubber.

3. **Hard-switching conditions**: This transition to hard-switching can introduce voltage spikes and increased EMI, which are undesirable in sensitive applications. Consequently, the effectiveness of the lossless snubber may diminish under conditions of high current, limiting their utility.
4. **Current limitations**: Generally, it is advisable to limit the maximum current handled by the power lossless snubber to around 50–70 A to avoid complications. When the current exceeds this threshold, careful attention is needed regarding PCB layout and the selection of components to ensure optimal performance and reliability.

3.2.3.2 Design Considerations

The effective implementation of lossless snubbers in DC–DC converters entails several key considerations:

1. **PCB layout**: A well-designed PCB layout is critical for minimizing stray inductance in the snubber circuit. Keeping traces short and wide helps reduce parasitic inductance, enabling faster response times and more effective snubber operation.
2. **Component selection**: Choosing the right components is equally important. The semiconductors used for the snubber should have packages with minimal bond wire inductance. The diode should feature a low forward voltage drop and sufficient current-handling capability to ensure efficient operation under the expected load conditions.
3. **Energy recovery**: Lossless snubbers are particularly effective on the high-voltage side of DC–DC converters, especially when commutation currents do

not exceed 100 A. They allow for energy recovery during switching events, thereby improving efficiency and reducing wasted energy as heat.

4. **Challenges on the low voltage side**: In contrast, the application of lossless snubbers on the low-voltage side, where commutation currents may exceed 200 A, presents significant challenges. In such scenarios, it is more prudent to design circuits where the commutation current remains below 50–70 A. This consideration helps prevent excessive voltage spikes and potential component failure.

3.2.3.3 Conclusion

In summary, lossless snubbers serve as powerful tools for managing switching behavior in DC–DC converters when applied correctly. They facilitate energy recovery and help mitigate the adverse effects of voltage spikes and EMI, particularly in high-voltage applications. However, careful design considerations, including PCB layout optimization and component selection, are vital to ensuring reliable operation, especially in high-current environments.

3.3 Operating Modes in Power Converters: Constant Voltage, Current, and Power

The process of power conversion is facilitated by power converters. As mentioned earlier, the primary goal of a power converter is to efficiently transfer energy from the power source to the load. These converters can operate in three main modes: constant voltage, constant current, and constant power. The mode in which a converter operates depends on the load conditions and the behavior of the control loops. The response of the control loops determines the converter's mode of operation. For example, a converter may switch from constant voltage to constant current mode based on feedback from the control loops, indicating a significant drop in load. Alternatively, this transition may be influenced by the parameters of the power stage, such as a dosed energy transfer configuration.

3.3.1 Constant Voltage Converter

Constant voltage converters are among the most commonly used types of power converters, found in a wide range of applications. They are designed to maintain a constant output voltage, regardless of changes in load or input voltage. This functionality is critical in powering sensitive electronic devices, where voltage stability is essential to ensure proper operation and prevent damage.

3.3.1.1 Operation and Control

A constant voltage converter typically uses feedback control circuits to achieve voltage regulation. The voltage control loop monitors the output voltage and dynamically adjusts the duty cycle in a PWM-controlled converter to compensate for fluctuations in the input voltage or load conditions. For example:

- If the load current increases, the control loop responds by increasing the duty cycle, thereby delivering more energy to maintain a stable output voltage.
- Conversely, if the load current decreases, the duty cycle is reduced to prevent overvoltage.

This regulation ensures that the output voltage remains constant or deviates only minimally, typically within between ±0.1% and ±1% of the nominal value. Such precise control results in a low internal output impedance, often in the micro-ohm to milliohm range, which makes these converters highly effective in voltage-sensitive applications.

3.3.1.2 Challenges in Parallel Operation

One of the challenges with constant voltage converters is managing parallel connections. When multiple constant voltage sources are connected in parallel, even slight differences in output voltage can lead to significantly different current levels between the converters. To address this, additional current-sharing control mechanisms are required. Without such measures, the imbalance can cause overheating, efficiency loss, or damage to the system. In real-world applications, batteries serve as an excellent example of constant voltage sources. However, connecting batteries in parallel, especially lithium-ion batteries, can pose risks:

- Variations in battery voltage can cause uneven current sharing, leading to overcharging or deep discharge in individual batteries.
- Differences in internal impedance may exacerbate the imbalance, reducing the overall lifespan of the battery pack.

3.3.1.3 Applications

Constant voltage converters are widely used in:

1. Consumer electronics, such as power adapters for laptops and phones
2. Industrial applications, such as powering control circuits and sensors
3. Battery chargers during the final phase of charging, where precise voltage regulation is critical

3.3.2 Constant Current Converter

A constant current converter is designed to maintain a stable output current, regardless of variations in load resistance or input voltage. This is particularly useful in applications where a consistent current is critical to the performance and longevity of the load.

3.3.2.1 Operation and Control

Constant-current operation is typically achieved through feedback control. The converter monitors the output current and dynamically adjusts the output voltage to ensure the desired current level is maintained. For instance:

- If the load resistance increases, the converter compensates by raising the output voltage to maintain the current.
- Conversely, if the load resistance decreases, the converter reduces the output voltage to prevent overcurrent conditions.

Example: LED Drivers

A common application of constant current converters is in LED drivers. LEDs are highly sensitive to current variations; even slight overcurrent conditions can cause overheating or premature failure. Constant current LED drivers regulate the current flowing through the LEDs, ensuring consistent brightness and protecting the components. These drivers automatically adjust their output voltage to match the load while maintaining a constant current.

Theoretical Impedance and Practical Considerations

In theory, the internal impedance of a current source is infinite, meaning that it can maintain the same current regardless of the load voltage. However, this characteristic introduces certain challenges:

- **Disconnected load**: If the load is disconnected or its resistance drops significantly, the output voltage can rise to extremely high levels. This overvoltage can damage connected components or the converter itself unless protective measures (e.g., overvoltage clamps) are implemented.
- **Parallel operation**: Unlike constant voltage converters, constant current converters can be connected in parallel without limits on the number of converters. This is because there are no issues with current sharing, as the output current of each converter remains independent.

Applications

Constant current converters are widely used in various applications, including:

- **LED lighting systems**: To ensure uniform brightness and prevent damage to LEDs.
- **Laser diode drivers**: For precise control of current in laser systems.
- **Battery charging circuits**: For certain battery chemistries, such as lithium-ion, during the initial charging phase where current regulation is crucial.

Limitations and Practical Examples

In real-life scenarios, generators are often modeled as current sources with voltage control loops. However, generators are not ideal for loads with large, abrupt changes in demand, as this can lead to instability or inefficient operation.

When using constant current converters, careful attention must be given to load characteristics and protective mechanisms to avoid issues like overvoltage under no-load conditions. These considerations make constant current converters indispensable in specialized applications where current regulation is more critical than voltage stability.

3.3.3 Constant Power Converter

A constant power converter is designed to provide a stable output power by dynamically adjusting the output voltage and current. It ensures that the product of voltage (V) and current (I) remains constant ($P = V \cdot I$), even when load conditions or input voltage vary. This capability makes constant power converters highly valuable in applications requiring precise power delivery and adaptability to changing operating conditions.

3.3.3.1 Operation and Control

The defining feature of a constant power converter is its ability to adjust its output characteristics:

- **If the output voltage decreases**, the output current increases proportionally to maintain constant power.
- **If the output voltage increases**, the output current decreases to keep the power output stable.

This behavior contrasts with constant voltage and constant current converters, as constant power converters adapt both voltage and current to meet the system's power demand.

3.3.4 Dosed Energy Transfer

An example of constant power implementation is the **dosed energy transfer** method. In this approach:

- Energy is transferred from the source to the load in discrete "doses."
- Each dose represents the energy stored and released by a capacitor or another energy storage element.

The power transferred can be described by the equation:

$$P = V^2 \cdot C \cdot f_{COM}, \tag{3.1}$$

where

- V is the voltage that charges and discharges the capacitor
- C is the capacitance
- f_{COM} is the commutation frequency

This method is commonly used in **resonant topologies**, which rely on resonant components to achieve efficient energy transfer. For instance, basic resonant topologies often include clamp diodes connected in parallel with resonant capacitors to control the flow of energy and ensure efficient operation.

3.3.4.1 Advantages

One key advantage of constant power converters is their ability to prevent the primary energy source—such as a grid or battery—from exceeding its maximum power delivery capability. This feature is especially critical in systems where the load demand fluctuates significantly. For example:

- As the load increases, the converter naturally transitions from voltage source mode to power source mode, driven by the parameters of the power stage circuit.
- This self-regulation makes the system more reliable without requiring additional external control.

3.3.4.2 Applications

Constant power converters are well-suited for applications where precise power delivery is essential or where the load characteristics vary significantly:

1. **Electric vehicle charging stations**: These converters optimize charging efficiency by adapting the charging rate based on the battery's state of charge.
2. **Capacitor charging**: High-value, high-voltage capacitors often rely on constant power converters for efficient and safe charging.

3. **Plasma loads and RF amplifiers**: These systems require stable power sources to operate effectively.
4. **Resonant topologies**: Used in scenarios where controlled energy transfer is critical.

Example: Electric Vehicle Charging

In electric vehicle charging stations, constant power converters maximize efficiency by transitioning seamlessly between different modes. For instance:

1. During the initial stages of charging, a high-power output is provided to quickly charge the battery.
2. As the battery approaches full charge, the converter adjusts to maintain optimal power delivery without overloading the battery.

Reliability Considerations

By adapting the output characteristics based on the load, constant power converters inherently offer improved reliability. This design philosophy ensures that the converter can handle dynamic load conditions without manual intervention, making it a preferred choice in modern power electronics design.

3.3.5 Summary

Understanding the distinctions between constant voltage, constant current, and constant power converters is essential for successful implementation in power electronics. Each converter type has its own set of advantages, applications, and requirements. A deep understanding of these factors, along with careful consideration of load characteristics and operational conditions, leads to more efficient, reliable, and optimized power conversion solutions.

Converters—especially those designed for higher power—must be capable of seamlessly transitioning between different modes, such as voltage source to current source, and vice versa, without introducing collateral effects. Incorporating the ability to switch between these modes enhances the versatility and performance of the converter. Additionally, adding the ability to transition to a constant power mode can further optimize converter operation in certain applications.

For example, in a battery charging application, a constant current converter is often used initially to rapidly charge the battery. Once the battery reaches a certain state of charge, the converter switches to a constant voltage mode to prevent overcharging and maintain the full charge. In many cases, a combination of constant current and constant voltage modes is employed to balance charging speed with long-term battery health and efficiency.

A key note regarding isolation DC–DC converters: capacitors, which are typically regulated by the voltage, function as voltage sources, while inductors are governed by current and serve as current sources. This understanding is critical when designing converters with strict isolation and precise control over voltage and current delivery.

By thoughtfully considering the operational modes and characteristics of power converters, engineers can design systems that are more adaptable, efficient, and capable of meeting the demands of modern power electronics applications.

3.4 Conclusion

This chapter explored the principles of isolated DC–DC converters. It begins with a set of aspirational criteria for power conversion hardware: 100% efficiency and continuous power flow from input to output. The authors recognize that achieving these ideals is impossible, so we present tools and methods that design engineers can use to approach the concept of an "ideal" power converter.

The chapter begins by discussing methods to control power devices, focusing on two approaches: (1) hard-switching and (2) soft-switching. Hard-switching is easier to implement than soft-switching but has significant drawbacks, such as higher electromagnetic interference (EMI) and reduced component lifetime. For example, the increased EMI often requires design engineers to use more complex filtering to meet specifications. Soft-switching, by contrast, is more challenging to implement but minimizes EMI by controlling slew rates. This typically results in a design that is both cost-effective and reliable.

With these considerations in mind, the chapter reviews zero-voltage switching (ZVS) and zero-current switching (ZCS). It introduces a lossless snubber to address shortcomings in some of the converter topologies discussed in Chap. 6. The chapter also examines the different operating modes of a converter, including constant voltage, constant current, and constant power modes. Finally, it describes Dosed Energy Transfer, which is integral to the implementation of resonant converters. By using these tools, the design engineer can approach the ideal power converter for their application.

3.5 Supplemental

The content of this chapter represents the original, creative work of the authors. We utilized OpenAI's ChatGPT to check grammar and spelling and to improve the clarity of certain portions of the text. Some of the images and data contained in this chapter were provided by Havis, Inc. with support from Alec Cook. Portions of the text and images are from tutorials that we presented at the *2017 IEEE Energy Conversion Congress and Exposition (ECCE 2017)*, Cincinnati, OH, USA, 2017 and *2022 IEEE Energy Conversion Congress and Exposition (ECCE 2022)*, Detroit, MI, USA, 2022.

Chapter 4
Semiconductor Devices

4.1 Silicon Power Devices

Silicon (Si) has been the dominant material for power semiconductors for decades due to its excellent balance of properties, including high carrier mobility, low cost, and well-established manufacturing processes. However, as power electronics evolve, silicon devices face challenges, particularly in high-temperature and high-frequency applications. The two main types of Si-based power devices are IGBTs and MOSFETs.

4.1.1 IGBTs (Insulated Gate Bipolar Transistors)

IGBTs combine the gate-drive simplicity of MOSFETs with the current-handling capability of bipolar transistors. This hybrid design allows for high voltage and current handling, making them ideal for applications requiring robust power delivery, such as industrial motor drives and high-voltage DC–DC converters.

However, IGBTs have slower switching speeds compared to Si MOSFETs, leading to higher switching losses in high-frequency environments. This drawback becomes significant when the commutation frequency exceeds 20 kHz, where efficiency is critical. Around two decades ago, IGBTs were often used in isolated DC–DC converters with commutation frequencies near 100 kHz, but their limitations in high-speed switching have since led to a decline in their use for this purpose.

Today, while IGBTs remain a reliable choice for motor drives and other high-voltage, low-frequency applications, they are largely replaced by faster alternatives like Si MOSFETs and wide-bandgap semiconductors—silicon carbide (SiC) and gallium nitride (GaN)—in high-frequency DC–DC converters. Therefore, this book

A. Isurin, M. J. Scott, *Practical Design Considerations for Isolated DC-DC Converters*, https://doi.org/10.1007/978-3-031-94427-7_4

will focus more on MOSFETs and emerging technologies, with minimal further discussion on IGBTs.

4.1.2 Si MOSFETs (Silicon Metal-Oxide-Semiconductor Field-Effect Transistors)

Si MOSFETs operate purely as field-effect transistors, relying on gate voltage to control the flow of current between the source and drain. They are known for their fast-switching speeds and low-switching losses, making them ideal for high-frequency applications such as DC–DC converters, inverters, and switch-mode power supplies (SMPS).

While their voltage ratings typically extend up to around 1.2 kV, they excel in medium-voltage and high-current applications. Their high current-handling capability makes them particularly suitable for low-voltage systems (up to 200 V) where high current, even exceeding 3000 A, can be achieved by connecting multiple MOSFETs in parallel. This scalability has made Si MOSFETs the preferred choice for low-voltage, high-current designs due to their cost-effectiveness and superior performance.

Another key advantage of Si MOSFETs is the wide variety of package options available. From surface-mount packages for compact designs to larger through-hole packages for higher thermal dissipation, engineers can select the most appropriate package to meet specific design and thermal management requirements. This versatility further enhances their appeal in diverse applications ranging from consumer electronics to automotive power systems.

4.1.2.1 Challenges and Considerations in Parallel MOSFET Design

When connecting Si MOSFETs in parallel to handle extremely high current (e.g., 3000 A or more), several key challenges must be addressed to ensure reliable and efficient operation. Below are the most critical factors design engineers need to consider:

1. **Current sharing**: MOSFETs connected in parallel do not always share current equally, which can lead to one or more devices carrying excessive current, overheating, and possibly failing. Several factors influence current sharing:
   - **On-resistance (R_{DS_ON}) variations**: Even slight differences in the MOSFETs' on-state resistance can cause uneven current distribution. Using devices with closely matched R_{DS_ON} specifications can mitigate this issue.
   - **Gate drive synchronization**: Inconsistent gate drive signals can result in different switching times for each MOSFET. A well-designed gate drive circuit

with low propagation delay and precise signal timing ensures synchronized switching.

2. **Gate drive design**: Driving multiple MOSFETs in parallel requires careful gate drive circuit design to ensure consistent and fast switching. Key considerations include:

   - **Gate resistors**: Individual gate resistors for each MOSFET can help dampen oscillations and ensure uniform switching speeds.
   - **Gate charge (Q_G) requirements**: Paralleled MOSFETs increase the total gate charge (Q_G), requiring stronger gate drivers capable of sourcing and sinking higher currents to maintain fast switching.
   - **PCB layout**: Minimizing parasitic inductance in the gate drive path is essential to prevent oscillations or switching delays. Using short, wide traces and Kelvin-source connections can help reduce these parasitics.

3. **Thermal management**: Handling high current generates significant heat, making effective thermal management a priority:

   - **Heatsinks and thermal pads**: Adequately sized heatsinks and thermal pads ensure efficient heat dissipation from each MOSFET. For high-power designs, forced-air cooling or liquid cooling systems may be necessary.
   - **Thermal interface materials (TIM)**: Using high-performance TIMs between the MOSFETs and heatsinks improves thermal conductivity and reduces temperature gradients.

Conclusion

Paralleling Si MOSFETs offers a cost-effective solution for achieving extremely high current in low-voltage applications. However, it requires careful consideration of current sharing, gate drive design, thermal management, and parasitic effects. By addressing these challenges, engineers can ensure reliable and efficient performance in demanding high-current systems.

4.1.3 Wide Bandgap (WBG) Semiconductors

The other group of semiconductors is Wide Bandgap (WBG) devices, which are semiconductor components that use materials with wider bandgaps than traditional Si. These materials include silicon carbide (SiC) and gallium nitride (GaN), and they offer several advantages over Si-based devices, particularly in power electronics applications [1–4].

WBG materials allow for faster switching speeds and lower switching losses, leading to improved efficiency. Lower conduction and switching losses result in higher efficiency, which is crucial for energy-sensitive applications [5–7].

- SiC transistors are suitable for the high-voltage side of isolated DC–DC converters, providing reliable performance in high-power systems.
- GaN transistors are suitable for the high-voltage side up to around 500 V and for the low-voltage side of isolated DC–DC converters without significant limitations.

These advantages make WBG devices increasingly popular in modern power electronics, particularly for demanding applications requiring higher efficiency and power density.

Design Trade-Offs

While WBG devices offer exceptional performance, they require careful consideration in terms of gate drive design, EMI management [8, 9], and thermal control due to their fast-switching speeds and higher voltage transients. Designers must balance these challenges against the benefits of reduced system size and improved efficiency. Connecting GaN devices in parallel demands more attention compared to Si, because GaN transistors are much faster than Si and have a smaller capability to withstand high current stress for short periods of time.

4.1.3.1 Highlighting Key Challenges and Trade-Offs When Using GaN Devices

GaN transistors offer impressive performance benefits, including faster switching speeds, lower conduction losses, and higher power density. However, these advantages come with specific design challenges that require careful consideration to ensure reliable and efficient operation.

1. **Gate drive design**: GaN transistors require precise gate drive control due to their fast-switching characteristics. Their low gate charge and narrow gate voltage range increase the risk of overshoot and ringing, potentially leading to device damage or unreliable switching. Specialized gate drivers with low propagation delay and strong gate damping are essential to manage these risks effectively.
2. **EMI management**: The high *dV/dt* and *dI/dt* associated with GaN transistors can lead to significant electromagnetic interference (EMI) [9]. Designers must focus on minimizing parasitic inductance through careful PCB layout, implementing proper decoupling strategies, and using effective shielding techniques to control EMI without compromising system performance.
3. **Thermal control**: Although GaN devices are more efficient and generate less heat under nominal conditions, their smaller die size and higher power density can lead to concentrated thermal hotspots. Effective heat dissipation methods, such as advanced thermal interface materials, optimized heat sinks, and innovative packaging, are crucial to prevent thermal stress and enhance long-term reliability.
4. **Paralleling challenges**: Paralleling GaN transistors presents additional complexities compared to Si-based devices. Due to their rapid switching and low

internal resistance, GaN devices are more sensitive to current imbalance, which can result in uneven current sharing and localized heating. Addressing these issues requires precise gate drive synchronization, minimal layout parasitics, and current balancing techniques to ensure reliable parallel operation.

In summary, while GaN transistors offer remarkable advantages in terms of switching speed, efficiency, and system miniaturization, their successful implementation necessitates a careful balance of gate drive design, EMI control, thermal management, and paralleling strategies. Properly addressing these trade-offs is essential for leveraging the full potential of GaN technology in modern power electronics applications.

4.1.3.2 Reliability and Longevity

A key characteristic of WBG semiconductors is that their die size is typically around five times smaller than a Si MOSFET's when comparing transistors with the same current and voltage. While this miniaturization offers benefits such as reduced parasitic losses and higher switching speeds, it also introduces significant challenges in thermal management, particularly under transient stress conditions.

Thermal excursion in semiconductors refers to the rapid temperature rise that occurs when the device's thermal management system is unable to dissipate heat effectively. This can lead to overheating, resulting in temporary performance degradation or, in severe cases, permanent damage due to thermal runaway. The reduced die size exacerbates this issue by limiting the surface area available for heat dissipation, making efficient cooling more challenging.

Additionally, WBG semiconductors, particularly their gate structures, are more sensitive to voltage fluctuations and stress than traditional silicon devices. This sensitivity increases the risk of gate oxide breakdown or threshold voltage shifts, which can further impact the device's reliability and longevity.

Given these factors, ensuring reliable operation and extending the lifespan of WBG devices requires meticulous attention to thermal design, gate drive protection, and overall system robustness. These considerations are critical in high-power applications, where thermal stress and electrical transients are common. Therefore, addressing these challenges is essential for maximizing the reliability and longevity of WBG semiconductors—a point we must emphasize once again.

4.1.3.3 Future Outlook of Wide Bandgap Devices

The future of power electronics is increasingly focused on the advancement of WBG materials. With ongoing research into ultra-wide bandgap materials like Gallium Oxide and Diamond, the next generation of power devices promises even greater efficiency, higher voltage tolerance, and reduced switching losses. These next-generation materials will likely outperform current WBG technologies, enabling even more compact and efficient power systems across a broader range of applications. As the technology matures, WBG semiconductors are poised to become the dominant materials in power electronics, revolutionizing industries from automotive to industrial automation.

4.2 Comparison of the Advantages and Disadvantages of Power Devices

As mentioned before, an isolated DC–DC converter consists of a high-voltage side and a low-voltage side. Let's start by considering which transistors we can use on the high-voltage side. We will compare Si and SiC devices. GaN will be discussed separately because it has specific gate drive requirements. WBG transistors have smaller switching times and significantly lower conductive losses. All these factors support the use of a commutation frequency of up to 1 MHz for high power, provided one knows how to implement it effectively. The primary goal to achieve good performance for both the low-voltage and high-voltage sides is to use soft-switching technology and increase the commutation frequency to the maximum extent possible. The results will improve even further when both strategies are combined.

Soft-switching technology can be used at any commutation frequency; however, the commutation frequency itself is functionally limited. Ten to fifteen years ago, the primary limiting factor on commutation frequency for high-power converters was the high-voltage side (above 400 VDC). This limitation was dictated by the characteristics of semiconductors such as IGBTs and MOSFETs, where a reasonable frequency was capped at around 150 kHz. When WBG (SiC) transistors first entered the market, their high cost made them impractical for automotive applications. Currently, the cost of SiC has decreased significantly, making them a viable option for widespread use.

Preliminarily, SiC is superior to Si transistors, but this does not imply that Si can be directly replaced by SiC; there are numerous differences between the two. A key distinction is that the die size of SiC is approximately five times smaller than that of Si when comparing transistors with the same current and voltage. Furthermore, SiC transistors exhibit relatively higher thermal resistance (°C/W), indicating that SiC is more sensitive to stress than Si.

Currently, many SiC transistors are available on the market, suitable for the high-voltage side of DC–DC converters. There are numerous application notes to aid in selecting a manufacturer. We will present a practical comparison based on the authors' experiences with a particular project, which aims to compare Si and SiC transistors and assess the feasibility of replacing Si with SiC. The data comes from a project where we retrofitted power conversion equipment for a hybrid electric bus. Notably, both devices utilized ISOTOP packages, allowing for a drop-in replacement.

Let's consider IXFN32N120P and MSC040SMA120J. Table 4.1 illustrates the main parameters of these transistors, indicating that SiC outperforms Si. Next, Fig. 4.1 displays the *I–V* characteristics of the transistors at $T_J = 25$ °C. Here, a significant difference is observed: SiC requires a higher gate voltage for complete

Table 4.1 The main parameters of the transistor comparison (cost estimates are based on numbers from 2024)

	IXFN32N120P	MSC040SMA120J
Type	Si MOSFET	SiC MOSFET
Cost	\$62.45 (100 units)	\$33.77 (100 units)
Package	SOT-227	SOT-227
Voltage	1200 V	1200 V
Current (@ 25 °C)	32 A	53 A
R_{DS_ON}	310 mΩ (V_{GS} = 10 V)	50 mΩ (V_{GS} = 20 V)
Q_G	360 nC	137 nC
$V_{GS(th)}$	3.5 V (min)	1.8 V (min)
$V_{GS,MAX}$	±30 V	−10/+25 V

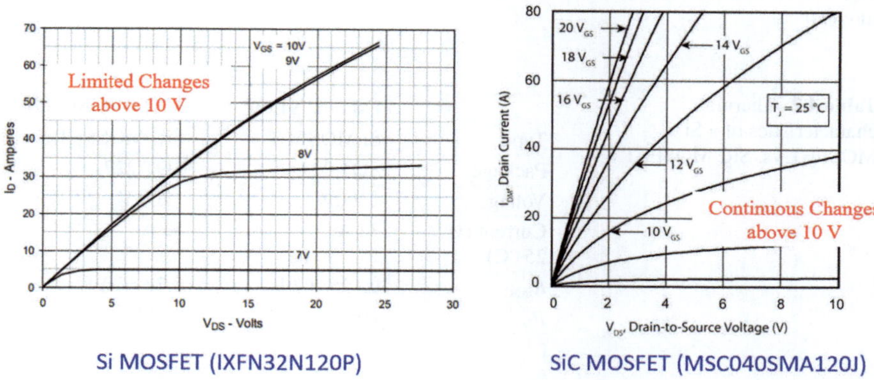

Fig. 4.1 *I–V* characteristics of an (**a**) Si MOSFET (IXFN32N120P) and (**b**) a SiC MOSFET (MSC040SMA120) at a junction temperature (T_J) 25 °C

turn-on, yet its maximum and transient voltages are lower—25 V for SiC and 30 V for Si.

This presents a major problem for hard switching:

- The gate voltage must transition from −4 V to approximately 20 V as quickly as possible to avoid excessive switching losses, necessitating a rapid rise from its minimum to its maximum gate voltage.
- This limits the maximum switching frequency of SiC devices.
- It negates some of the advantages of SiC devices.

Figure 4.2 illustrates that the normalized on-resistance (R_{DS_ON}) of Si MOSFETs varies more with temperature than that of SiC. For Si, moving from +25 to +125 °C results in a twofold increase, while below +25 °C, the R_{DS_ON} decreases by 0.6 times from +25 to +40 V. SiC, on the other hand, changes by only 1.3 times from +25 to +125 °C, and the R_{DS_ON} rises when temperatures fall below +25 °C.

Thus, Si MOSFETs are:

- Better suited for paralleling devices
- Easier to use with desaturation protection

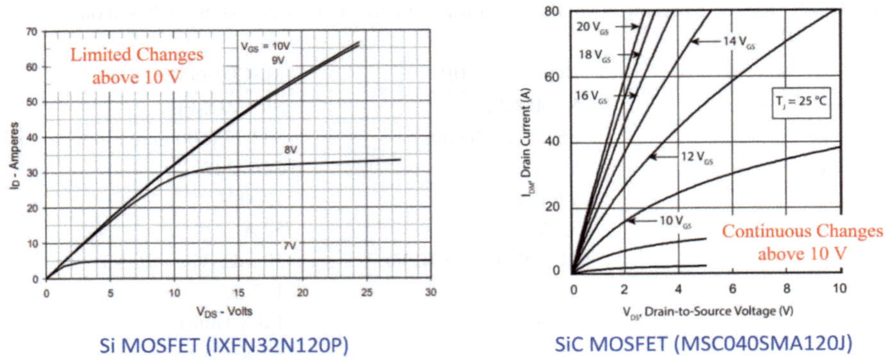

Fig. 4.2 Normalized R_{DS_ON} for a Si MOSFET and a SiC MOSFET versus the junction temperature (T_J)

Table 4.2 Thermal characteristics of a Si MOSFET vs. SiC MOSFET

	IXFN32N120P	MSC040SMA120J
Type	Si MOSFET	SiC MOSFET
Package	SOT-227	SOT-227
Voltage	1200 V	1200 V
Current (@ 25 °C)	32 A	53 A
$R_{th,JC}$	0.125 °C/W	0.72 °C/W
P_D	1000 W	208 W

These parameters require careful consideration, especially when connecting SiC transistors in parallel. Table 4.2 shows the main thermal characteristics ($R_{th,JC}$) and maximum power dissipation (P_D). While SiC transistors exhibit only conductive losses and clearly outperform Si in that regard, the question of overall performance, once switching losses are factored in, remains complex. Although the die of a SiC transistor is significantly smaller—around five times smaller than that of a Si transistor.

We have examined straight SiC, as well as hybrid SiC with a cascode structure. Figure 4.3 illustrates the configuration of a cascode device, which combines a normally-off, low-voltage Si MOSFET with a normally-on, high-voltage SiC JFET.

4.2.1 Operation

The MOSFET's drain-source voltage ($V_{DS,M}$) establishes a negative gate-source voltage ($V_{GS,J}$) for the SiC JFET with a value of $V_{GS,J} = -V_{DS,M}$ [10].

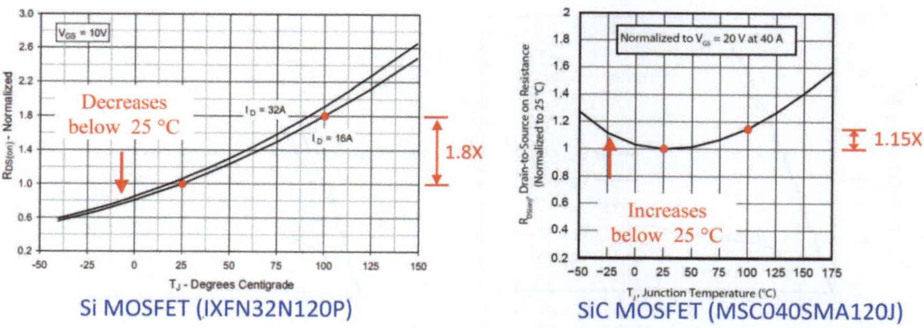

Fig. 4.3 Structure of a MOSFET and JFET in a cascode power device with a kelvin connection for the source

Table 4.3 Comparison of three types of transistors: a Si MOSFET, a SiC MOSFET, and a SiC cascode device

	IXFN32N120P	MSC040SMA120J	UF4C120053K4S
Type	Si MOSFET	SiC MOSFET	SiC cascode transistor
Cost	\$62.45[a] (100 units)	\$33.77[a] (100 units)	\$13.37[a] (100 units)
Package	SOT-227	SOT-227	TO-247-4
Voltage	1200 V	1200 V	1200 V
Current (@ 25 °C)	32 A	53 A	34 A
R_{RD_ON}	310 mΩ (V_{GS} = 10 V)	50 mΩ (V_{GS} = 20 V)	53 mΩ (V_{GS} = 12 V)
Q_G	360 nC	137 nC	37.8 nC
$V_{GS(th)}$	3.5 V (min)	1.8 V (min)	4 V (min)
$V_{GS,MAX}$	±30 V	−10/+25 V	±20 V
V_{SD}	1.5 V	3.9 V	1.28 V

[a]cost estimates based on 2024 prices

4.2.2 Benefits of Cascode Configuration

- Behaves like a Si MOSFET while leveraging SiC advantages
- Wider margin between V_{TH} and $V_{GS,MAX}$
- Smaller gate charge (Q_G)

Table 4.3 compares three types of transistors: Si MOSFET, pure SiC MOSFET, and a SiC cascode transistor. It's notable that there is significant cost differentiation between Si, SiC, and cascode configurations, with cascade transistors typically utilizing a TO-247 package. Figure 4.4 shows the comparison of *I–V* characteristics at T_J = 25 °C between pure SiC and cascade SiC. As we can see, the *I–V* characteristics of the SiC cascode are closer to that of Si MOSFETs. Figure 4.5 compares the normalized on-resistance (R_{DS_ON}) between SiC MOSFETs and cascode SiC transistors, revealing that the temperature dependency of R_{DS_ON} for the SiC cascode is similar to that of Si MOSFETs.

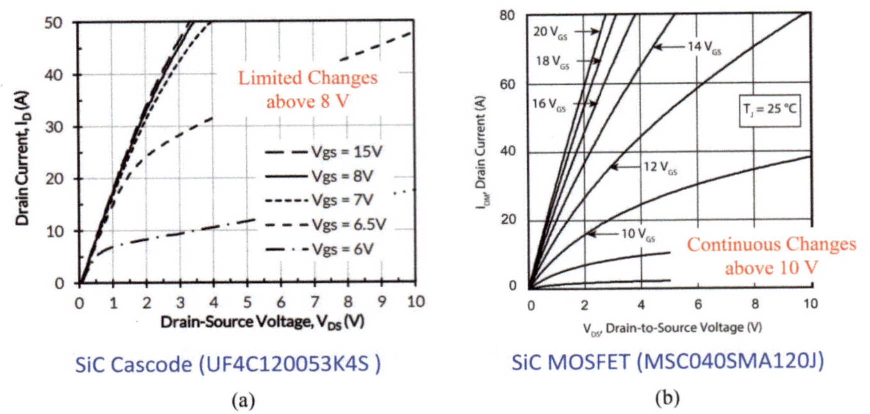

Fig. 4.4 Comparison of *I–V* characteristics between (**a**) cascode SiC and a (**b**) pure SiC MOSFET at $T_J = 25$ °C

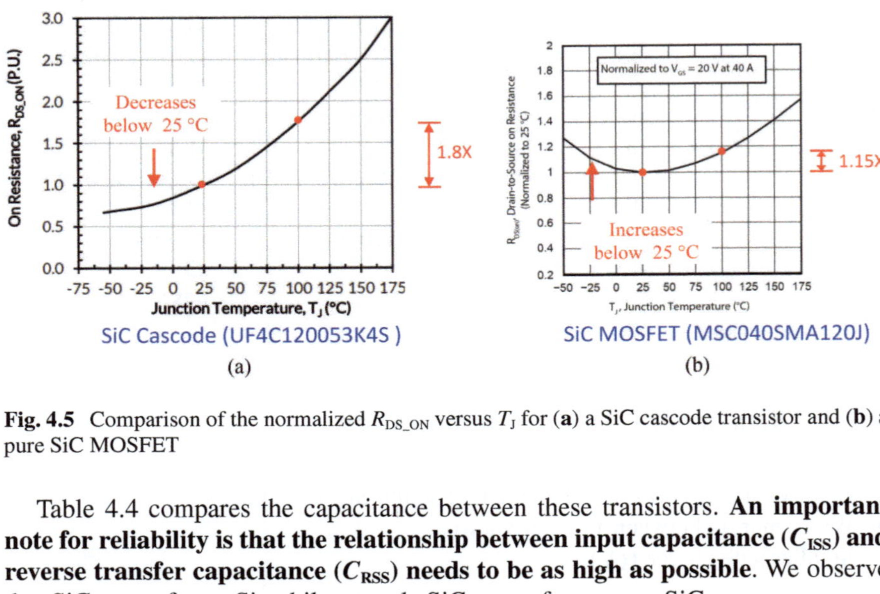

Fig. 4.5 Comparison of the normalized R_{DS_ON} versus T_J for (**a**) a SiC cascode transistor and (**b**) a pure SiC MOSFET

Table 4.4 compares the capacitance between these transistors. **An important note for reliability is that the relationship between input capacitance (C_{ISS}) and reverse transfer capacitance (C_{RSS}) needs to be as high as possible**. We observe that SiC outperforms Si, while cascade SiC outperforms pure SiC.

Historically, in the era dominated by bipolar transistors, cascode structures (from common emitter to common base) were used for high frequency and high voltage applications. During the development of new converters—especially when considering new transistors for the high-voltage side—engineers are always advised to conduct tests on the transistors. Double Pulse Testing for power semiconductor devices, using an oscilloscope and arbitrary function generator, as described in [6–8, 11], is one such method that reflects the specific application area of the transistors. This type of testing allows design engineers to observe the real behavior of

Table 4.4 The capacitance comparison among these transistors

$V_{CE} = V_{DS} = 25$ V	IXFN32N120P (IXYS)	MSC040SMA120J (Microsemi)	UF4C120053K4S (Qorvo)
C_{ISS}	21,000 pF	1990 pF	1370 pF
C_{RSS}	77 pF	17 pF	2.2 pF
C_{GS}/C_{DG}	271.7	117	622

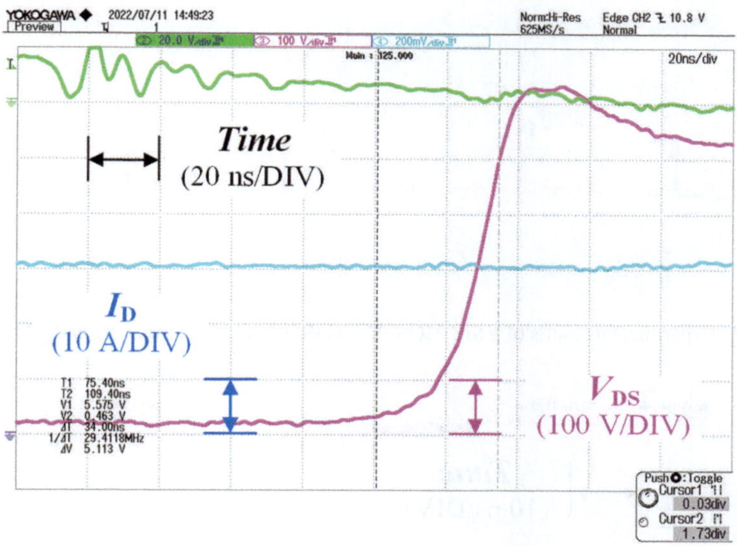

Fig. 4.6 Experimental results of a Si MOSFET (IXFN32N120P) turning off in 34 ns

transistors and predict potential issues during development. In other words, the time invested in testing can save time and lead to better product outcomes.

Figures 4.6, 4.7, and 4.8 demonstrate the test results obtained by the authors. The operating conditions were a device current of 30 ADC and a turn-off voltage of 500 VDC. Figure 4.6 shows the turn-off behavior of a Si MOSFET of 34 ns. Figure 4.7 displays the turn-off behavior of a SiC MOSFET at 20.4 ns. Finally, Fig. 4.8 is the turn-off behavior of the SiC cascode transistor at 18 ns. For these sets of tests, the SiC devices' turn-off time was nearly ½ of the Si MOSFET. We performed all three tests on the PCBs used in the application.

Although the transistors are intended for operation up to 850 VDC, testing was conducted at 500 VDC. It is preferable to test the devices at around 850 VDC, but 500 VDC suffices for this application. This testing is classified as dynamic.

Additionally, a static test is highly recommended. This test involves connecting the gate driver to the transistors (or using a capacitor equivalent to the gate capacitor of the transistor) and applying a signal from a function generator to the gate driver to observe the waveform at the transistor gate. The frequency generated should align with the predicted frequency in the converter design, and the gate driver voltage

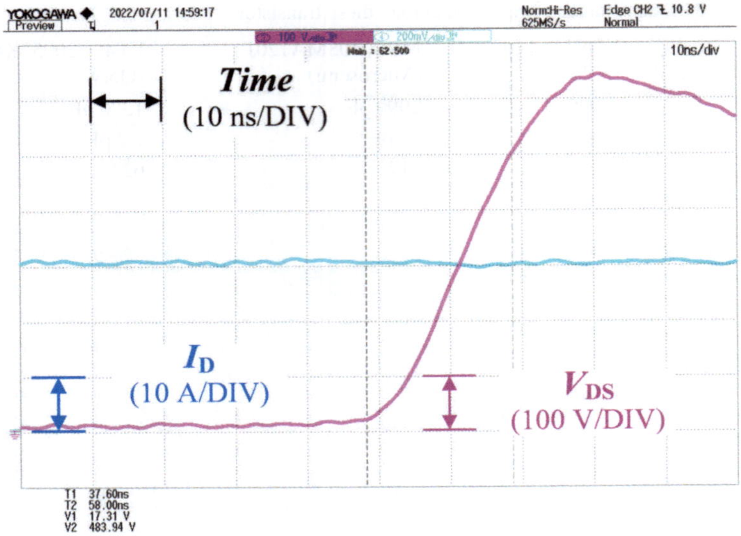

Fig. 4.7 Experimental results of a SiC MOSFET (G3R20MT12N) turning off in 20.4 ns

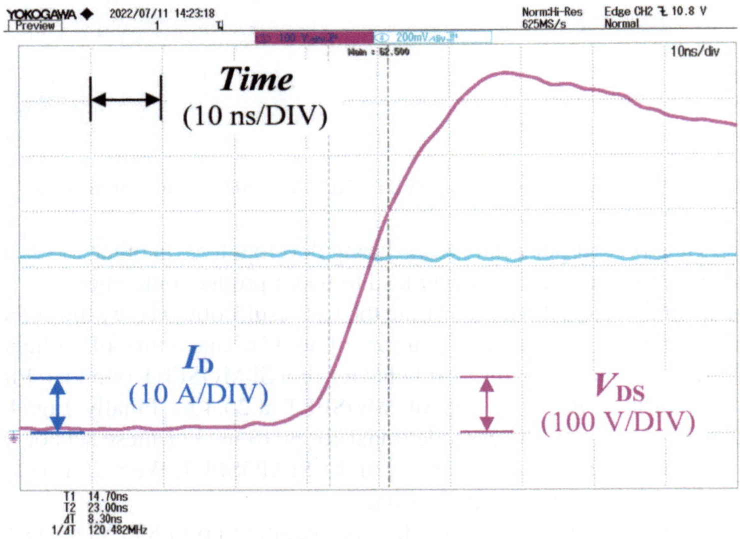

Fig. 4.8 Experimental results of a SiC cascode transistor (MSC040SMA120J) turning on in 8 ns

should match that of the intended operational voltage. Through this static test, design engineers can view the preliminary waveform and measure the power consumption of the gate driver. This static testing process should be conducted for both high-voltage and low-voltage side transistors and should be tested with a real PCB.

Now, let's turn our attention to GaN transistors for the high-voltage side. These transistors come in two configurations: pure GaN and cascade configuration. The cascade configuration is typically rated for high voltage, around 600 V, and is somewhat comparable to the cascode configuration of SiC transistors. The selection of transistors for the high-voltage side of a DC–DC converter ultimately depends on the design engineer's requirements.

Pure GaN exhibits unique features, with a low gate-to-source threshold voltage of 1.1–1.7 V and an absolute maximum of 6–7 V. The transient gate-to-source voltage is capped at 10 V. This characteristic necessitates significant attention from gate driver perspectives, particularly regarding immunity and reliability for the high-voltage side. In the gate driver circuits, the literature offers insights into what the authors implemented and recommended for addressing these challenges.

At present, many manufacturing companies produce a wide selection of transistors for the high-voltage side, and the choice will depend on the specific applications needed.

4.3 Mechanical and Electrical Challenges in Low-Voltage, High-Current Isolated DC–DC Converters

When designing isolated DC–DC converters, the low-voltage, high-current side presents unique challenges that require careful attention to transistor selection, mechanical implementation, power connections, PCB layout, and other factors. Let's begin by examining the specific issues related to this section. For achieving efficient power conversion, two primary approaches are employed: (1) soft-switching technology, which reduces switching losses and is adaptable to a wide range of switching frequencies, and (2) maximizing the switching frequency, which reduces the size of passive components and improves power density. However, practical limitations prevent reaching extremely high commutation frequencies. Previously, these limitations were predominantly found on the high-voltage side of the converter, but in modern systems, the low-voltage side has emerged as a new bottleneck.

4.3.1 Historical Limitations in Transistor Technologies

Fifteen to twenty years ago, the high-voltage side of DC–DC converters relied heavily on IGBTs and MOSFETs, each with distinct limitations:

1. **IGBTs**: Suffered from slower switching speeds, making them unsuitable for high-frequency operation
2. **MOSFETs**: Limited by the high gate-drive power requirements and cost, especially at high voltages above 400 VDC

For instance, driving three IXFN32N120P MOSFETs connected in parallel at 100 kHz required a gate driver power supply rated at 3 W. Achieving this while maintaining 2500 V isolation and keeping the input-to-output capacitance below 10 pF was a complex task. As a result, the practical maximum commutation frequency for high-voltage designs rarely exceeded 150 kHz. SiC components are emerging but remained prohibitively expensive during this period.

4.3.2 The Current Landscape: Shift in Limitations

Today, the landscape of power electronics has shifted significantly. SiC transistors are now widely available, offering improved performance at competitive prices. They enable high-voltage designs to achieve switching frequencies between 400 and 600 kHz. This advancement has alleviated the high-voltage side's frequency limitations, shifting attention to the low-voltage side, particularly in high-current applications.

4.3.2.1 New Limitations on the Low-Voltage Side

Two primary challenges arise when designing the low-voltage side of an isolated DC–DC converter operating at high currents: (1) mechanical constraints and (2) electrical limitations. Let's first examine the mechanical issues. As the current surpasses 300 A, the size of components associated with the low-voltage side—such as transistors, capacitors, busbars, and connections—must increase. This size increase directly impacts the PCB layout, introducing higher conduction losses and affecting overall efficiency. For instance, in a real-world DC–DC converter operating at 100 kHz with a 30 V, 500 A output (15 kW), the PCB's conduction losses on the low-voltage side reached approximately 300 W, representing 2% of the total power. Attempts to push the commutation frequency higher led to an unacceptable increase in conduction losses to 3.5%. Notably, this PCB was originally designed for 300 A, with conduction losses at 1% for that load. The issue arose from the decision to increase the output power without modifying the low-voltage PCB design, a practice that was sufficient when Si semiconductors were used on the high-voltage side, as boosting output power could often be achieved by simply swapping semiconductors without redesigning the entire PCB.

A similar limitation that was once associated with Si on the high-voltage side now affects the low-voltage side: the power supply requirements for gate drivers. For example, driving seven transistors per switch in a bridge rectifier for a 500 A output at 200 kHz requires approximately 5 W per gate driver, totaling around 20 W for the entire rectifier. Providing this power efficiently, while maintaining proper connections and isolation between the gate driver and transistors, is a complex task. Establishing stable, low-parasitic pathways in such high-current environments is not straightforward.

Ultimately, these challenges boil down to two key issues: the physical implementation of the low-voltage side and the power supply demands of the gate driver. The physical implementation relies heavily on the combined expertise of power electronics, mechanical, and PCB design engineers. The degree of success often hinges on how well these disciplines collaborate and whether they can integrate mechanical robustness with electrical efficiency. Secondly, budget constraints play a significant role. Using higher-cost components and advanced materials can yield significantly better results than sticking to cost-effective, lower-performance options. Numerous examples have demonstrated that investing in premium components and materials results in superior performance, lower losses, and increased reliability, particularly in high-power, high-current applications.

Another limitation related to the power supply for gate drivers can be mitigated by adopting energy recovery gate drivers. Various publications, including [12–15], discuss energy recovery gate drivers, each suited to specific technical requirements of different converter designs. The choice of the appropriate implementation depends on the particular converter topology and the design goals. In this context, the authors employed energy recovery gate drivers for both the high-voltage and low-voltage sides, as detailed in [16, 17].

The primary purpose of these gate drivers is to efficiently drive multiple transistors connected in parallel or handle transistors/modules with high gate capacitance. However, a common challenge arises from the internal gate resistors (R_I) that manufacturers include inside the packages. They diminish the effectiveness of energy recovery and ultimately limit the maximum achievable commutation frequency. Figure 4.9 illustrates the schematic of this energy recovery gate driver, highlighting the energy recovery transformer (W_1–W_2) as its core component. This transformer is essential for recycling energy from the gate drive circuit, resulting in a reduction of power consumption by 3–5 times compared to conventional gate drivers without energy recovery. The extent of this reduction depends on the implementation quality of the energy recovery transformer and the surrounding circuitry.

Another significant advantage of this energy recovery gate driver is its ability to mitigate transient overvoltage spikes by dynamically altering the effective gate inductance of the MOSFET. Figure 4.10 presents the equivalent schematic of the

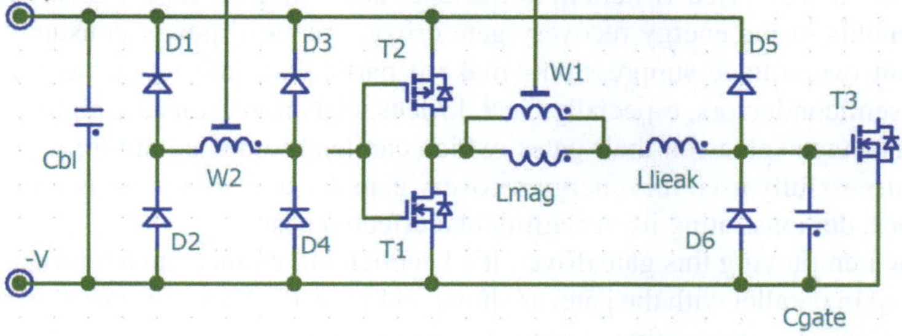

Fig. 4.9 Schematic of a gate drive circuit with energy recovery (U.S. Patent 6,570,416)

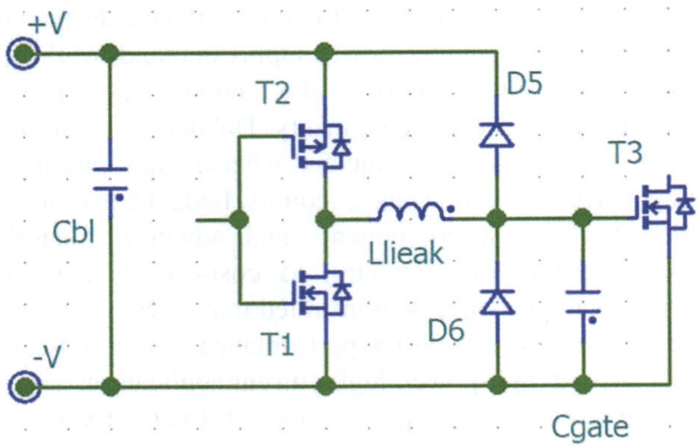

Fig. 4.10 The equivalent schematic of the gate driver during the charge or discharge phase, where L_{leak} and C_{gate} are resonance circuit

gate driver during the charge or discharge phase, where L_{leak} represents the leakage inductance of the energy recovery transformer. This inductance plays a crucial role in controlling the switching dynamics of the gate.

Figure 4.11 provides an alternative equivalent schematic, emphasizing the magnetizing inductance (L_{mag}) of the energy recovery transformer. The inductance in series with the transistor's gate capacitance (C_{gate}) is variable, ranging from 20–50 nH to 10–20 μH, depending on the switching state. This variability allows for optimized energy transfer and reduced switching losses.

Finally, Fig. 4.12 shows the experimental results for the gate drive signal under two static tests: (1) without the energy recovery transformer (Fig. 4.12a) and (2) one with the energy recovery transformer (Fig. 4.12b). The test utilized eight IRF3805 MOSFETs connected in parallel. The switching frequency was set to 200 kHz, with a gate drive voltage of approximately 12 V. With the energy recovery transformer, power consumption was reduced from 3 W (without the energy recovery transformer) to 1.2 W. Additionally, the overvoltage spikes in gate signal were eliminated. These conditions demonstrate the practicality and effectiveness of energy recovery in high-frequency, high-capacitance scenarios, highlighting the feasibility of achieving significant power savings while maintaining reliable gate drive performance even at elevated switching frequencies and high gate-capacitance. These two key benefits of the energy recovery gate driver—reduced power consumption and transient overvoltage suppression—make it particularly advantageous for driving WBG semiconductors, especially GaN devices. GaN transistors are highly sensitive to overvoltage spikes on their gates, which can lead to device failure. The authors have successfully used this energy recovery gate driver across various semiconductor types, demonstrating its versatility and effectiveness.

When employing this gate driver, it's beneficial to connect an additional capacitor (C_{add}) in parallel with the gate, as shown in Fig. 4.13. This configuration provides two distinct advantages: first, it reduces the gate's sensitivity to noise, which is critical for maintaining signal integrity in high-speed switching environments; second,

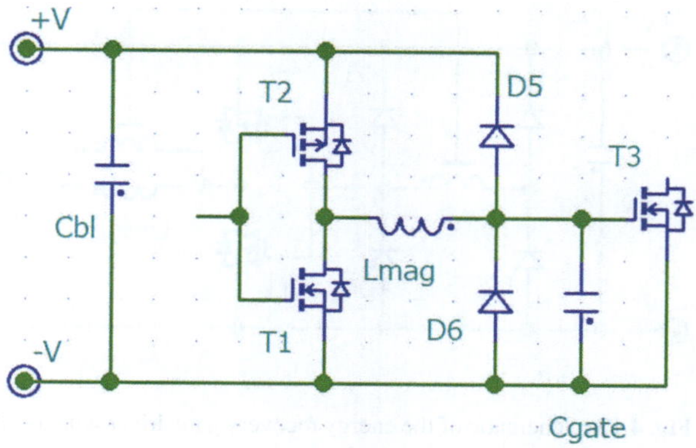

Fig. 4.11 The equivalent circuit of the gate driver during the charge or discharge phase, where L_{mag} and C_{gate} are the resonance circuit

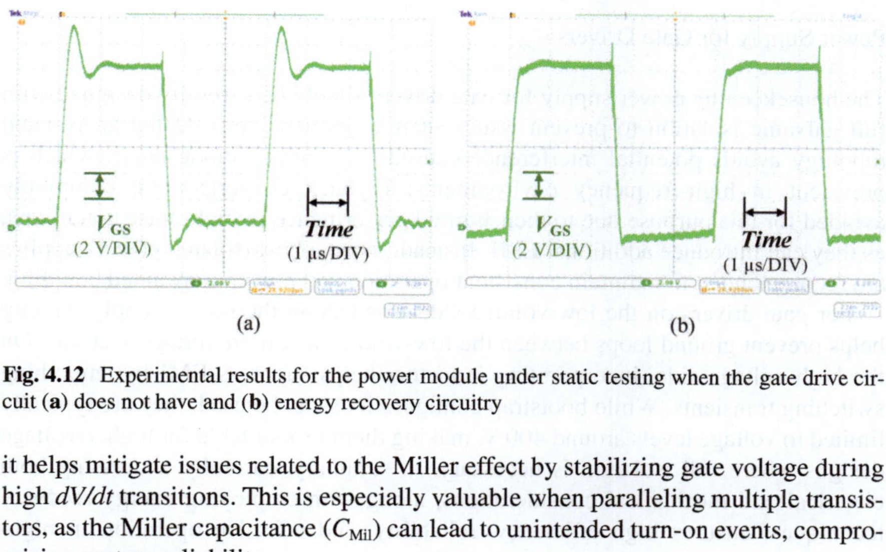

Fig. 4.12 Experimental results for the power module under static testing when the gate drive circuit (**a**) does not have and (**b**) energy recovery circuitry

it helps mitigate issues related to the Miller effect by stabilizing gate voltage during high *dV/dt* transitions. This is especially valuable when paralleling multiple transistors, as the Miller capacitance (C_{Mil}) can lead to unintended turn-on events, compromising system reliability.

In cases where 8–12 transistors are connected in parallel, current distribution among the transistors becomes even more critical, particularly with GaN devices due to their fast-switching characteristics. During ZCS during turn-on and turn-off, or when switching with a small current, it becomes essential to minimize the switching losses and prevent gate overstress. Additionally, it is advisable to centrally locate the gate driver connection relative to the transistors to ensure uniform current sharing and to reduce gate-drive loop inductance, which further mitigates ringing and oscillations. The resulting near-sinewave current profile at the gate reduces high-frequency noise emissions, which improves the system's EMI performance.

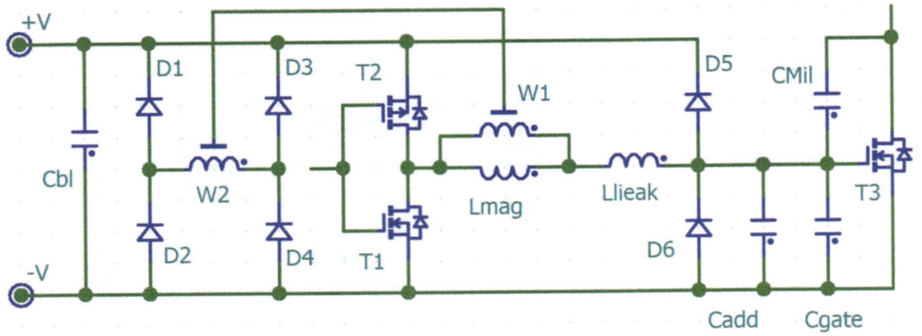

Fig. 4.13 Schematic of the energy recovery gate driver with an additional capacitor (C_{add}) in parallel with the gate-source terminals

Power Supply for Gate Drivers

The housekeeping power supply for gate drivers should be carefully designed with full galvanic isolation to prevent issues such as ground loops. Using an isolated topology avoids potential interference caused by common-mode noise, which is prevalent in high-frequency environments. Flyback converters are commonly avoided for this purpose due to their limited performance in high-noise conditions, as they can introduce additional EMI. Instead, more robust isolated power supplies are recommended to maintain consistent operation and prevent unwanted coupling.

For gate drivers on the low-voltage side, connecting the power supply directly helps prevent ground loops between the low-voltage and high-voltage sections. On the high-voltage side, incorporating inductors helps suppress EMI by smoothing switching transients. While bootstrap configurations can be used, they are typically limited to voltage levels around 400 V, making them less suitable for higher-voltage designs where full isolation is necessary. In summary, carefully addressing both electrical and mechanical considerations, along with employing energy recovery techniques, ensures high efficiency, reliability, and effective operation in high-current, high-frequency isolation DC–DC converters.

4.4 Ensuring Reliability in DC–DC Converters: Key Considerations and Practical Approaches

The reliability of power devices is a critical factor that directly influences their performance and longevity across various applications. It is evident that ensuring reliability is a fundamental task during the development process and in the long-term operation of power units. Reliability, by nature, is a complex and multifaceted subject involving various parameters such as thermal management, electromagnetic interference (EMI), component degradation, and failure modes.

In the context of isolation DC–DC converters, particularly on the high-voltage side, several key aspects must be considered to enhance reliability. These include the selection of appropriate semiconductors, such as Si MOSFETs or IGBTs or WBG devices. For example, GaN transistors offer superior switching characteristics and lower losses, but they require appropriate thermal management techniques to handle high power densities, and robust gate driver designs to ensure stable operation under high-voltage stress. Designers must also understanding the failure mechanisms related to insulation breakdown, voltage spikes, and thermal cycling. This is crucial for designing converters so they can withstand harsh operating conditions. In the following sections, we will explore these critical factors in more detail, focusing on practical strategies to improve the reliability of high-voltage DC–DC converters. Let's take a closer look at the transistor within its package. Figure 4.14 illustrates a typical transistor construction with Kelvin connections, which provides a dedicated path for the gate drive circuit to minimize the effects of common-source inductance ($L_{S,WB}$). However, bond wires inside the package introduce stray inductance ($L_{i,WB}$), which can lead to additional stress during high-speed switching. To mitigate voltage spikes at the gate-source terminals, manufacturers often integrate an internal gate resistor (R_I) as mentioned above. This resistor helps prevent excessive voltage overshoots but also limits the maximum switching speed and increases energy losses. In some cases,

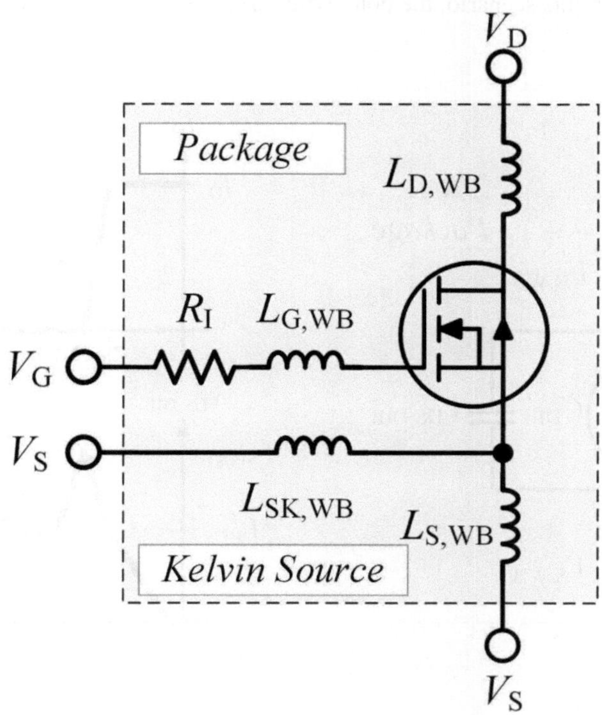

Fig. 4.14 Construction of a transistor with four-terminal Kelvin connections

the value of this internal resistor is so high that it significantly restricts the transistor's operational range, particularly in applications requiring fast switching transitions.

For this reason, it is preferable to use transistors with minimal internal gate resistance, especially in designs utilizing soft-switching techniques. Soft switching reduces the occurrence of voltage spikes, minimizing the stress on the gate without relying heavily on resistive damping. However, if a gate resistor is required for design stability or EMI control, it is advisable for design engineers to select an appropriate external resistor value to balance the switching speed, efficiency, and reliability of the power device.

In summary, understanding the internal and external gate resistances and their impact on switching behavior is crucial for optimizing transistor performance, particularly in high-efficiency DC–DC converter designs.

The design engineer must pay close attention to what happens inside the transistor package, even though it cannot be observed directly. Instead, its internal behavior can only be inferred through external measurements and predicted consequences—some of which may be severe or even catastrophic. Figure 4.15 illustrates one of these effects: the impact of commutation inductance. The stray inductance can cause overvoltage spikes (V_{SPIKE}) across the transistor die ($v_{DS,DIE}$), with the magnitude of the spike depending on how rapidly the transistor turns off. Additionally, the transistor experiences increased energy losses, which can shorten its lifespan. In this scenario, the bond wire serves as the primary energy source for the spike.

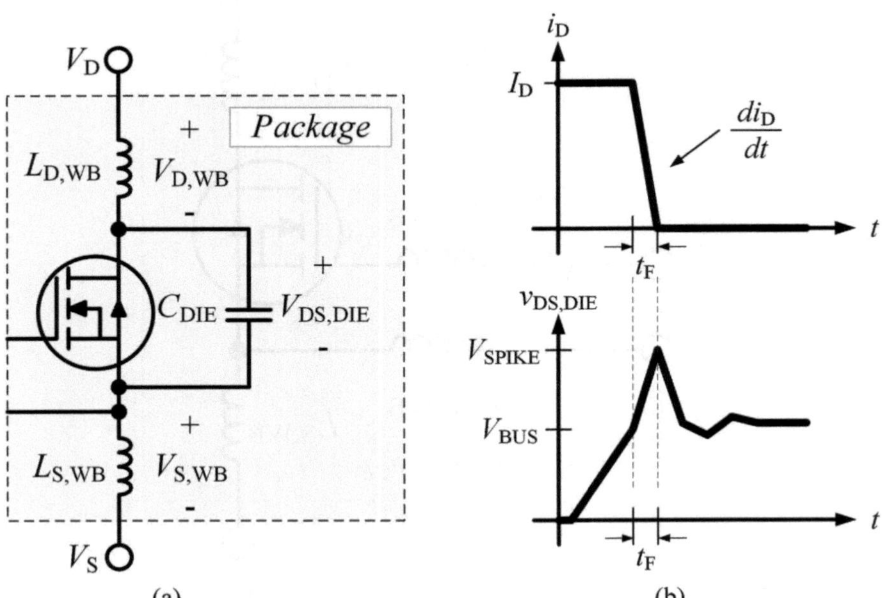

Fig. 4.15 (**a**) The schematic of a MOSFET with the parasitic components, LD,WB, CDIE, and LS,WB and (**b**) their effect on the MOSFET's behavior during turn-off event

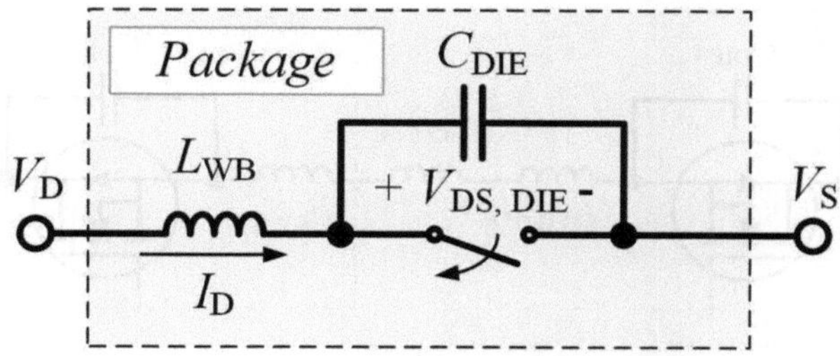

Fig. 4.16 The equivalent circuit of the MOSFET from Fig. 4.15 during a turn-off event

Figure 4.16 shows the equivalent circuit representing the non-ideal elements inside the package. This configuration resembles a resonant circuit formed by the bond wire inductance (L_{WB}) and the body capacitance of the transistor (C_{DIE}). The quality factor (Q-factor) of this circuit depends on the transistor's conductivity at the time of switching. WBG transistors, such as GaN or SiC devices, switch significantly faster than Si transistors, and this makes the issue more prominent. This explains why manufacturers often include an internal gate resistor (R_I) to dampen oscillations and mitigate overvoltage risks.

The problem becomes even more critical in AC commutation switches, where two transistors are connected in series. Figure 4.17 highlights this scenario. In such configurations, the stray inductance from both the trace connections (L_{CONN}) and wire bonds ($L_{S,WB}$ and $L_{D,WB}$) resonates with the die capacitance (C_{DIE}). The combined inductance from the bond wires and the trace connections results in a significantly higher energy buildup, amplifying overvoltage and energy losses compared to single-transistor setups. This excessive energy can lead to premature failure of the transistors.

Given these risks, it is crucial to design commutation circuits with extra care. To mitigate these effects, using a sinusoidal current waveform can improve performance. Additionally, employing pure ZVS techniques and increasing the internal turn-off time by adjusting the external gate resistor can further reduce overvoltage issues.

In Chap. 6, we will explore practical solutions for using these switches and discuss methods to minimize or eliminate the negative effects associated with commutation-induced overvoltages in real-world implementations.

Another critical issue with high *dV/dt* is its impact on the gate drives behavior, particularly through the Miller capacitance (C_{DG}). Most engineers are familiar with shoot-through conditions caused by current propagating through this capacitance. However, this challenge becomes even more pronounced when working with WBG devices due to their faster switching speeds and higher *dV/dt*.

This phenomenon occurs after the lower transistor turns off and while the upper transistor is turning on. During this switching transition, the voltage at the drain of the lower transistor rises rapidly. A portion of this voltage is dropped across C_{DG}, causing the current to flow through one of two paths: either through the gate resistor

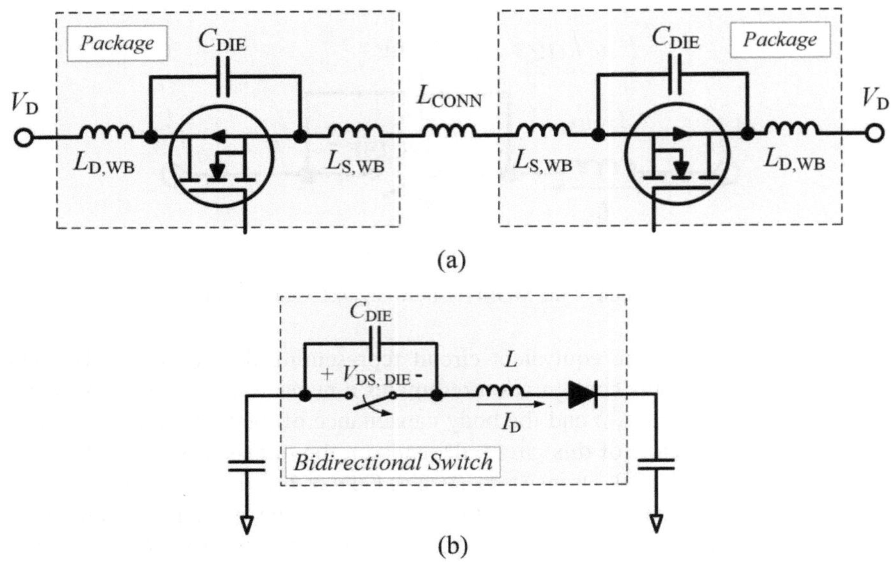

Fig. 4.17 (**a**) the schematic of a bidirectional switch with its parasitics elements and (**b**) its corresponding equivalent circuit

(R_{DR}) and gate inductance (L_{GR}) or through the gate-source capacitance (C_{GS}). This results in a momentary voltage spike at the gate.

If the *dV/dt*-induced current is too high, or if R_{DR} and L_{GR} are too large, or if the device's threshold voltage is low, this spike can become large enough to turn on the lower device while the upper device is also conducting. This creates a temporary shoot-through condition, leading to cross-conduction, which can result in device failure, tripping of protection circuits, or increased stress on components. In the long term, this issue can cause premature field failures, leading to high warranty costs for manufacturers.

Figure 4.18 illustrates the problem caused by Miller capacitance. To mitigate or significantly reduce this issue, an external capacitor (C_X) and resistor (R_X) can be added, as shown in Fig. 4.19. C_X provides an alternate current path, reducing the magnitude of the voltage spike (V_{SPIKE}) at the gate. Meanwhile, R_X lowers the *Q*-factor of the resonant circuit, damping oscillations and stabilizing the gate voltage. However, since C_X consumes additional energy from the gate driver, care must be taken to ensure the energy demand is not excessive. In high-power applications, a gate driver with energy recovery capability may be necessary.

The authors recommend always including an external capacitor to counteract the effects of Miller capacitance, thereby making the transistor's switching behavior more predictable and reliable. This approach reduces the risk of shoot-through, enhances system stability, and extends the overall lifespan of the power devices.

Another critical issue related to transistor packaging is the influence of common-source inductance (L_{CS}), as illustrated in Fig. 4.20. This problem arises when the

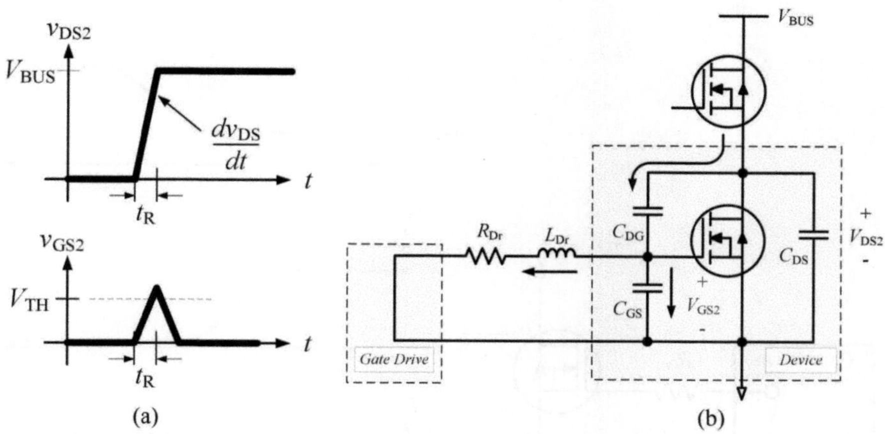

Fig. 4.18 The *Miller Capacitance* overshoot problem with (**a**) the waveforms for the drain-to-source voltage V_{DS2} and gate-to-source voltage V_{GS2} of the lower device, and (**b**) a schematic of a half-bridge converter with its parasitics

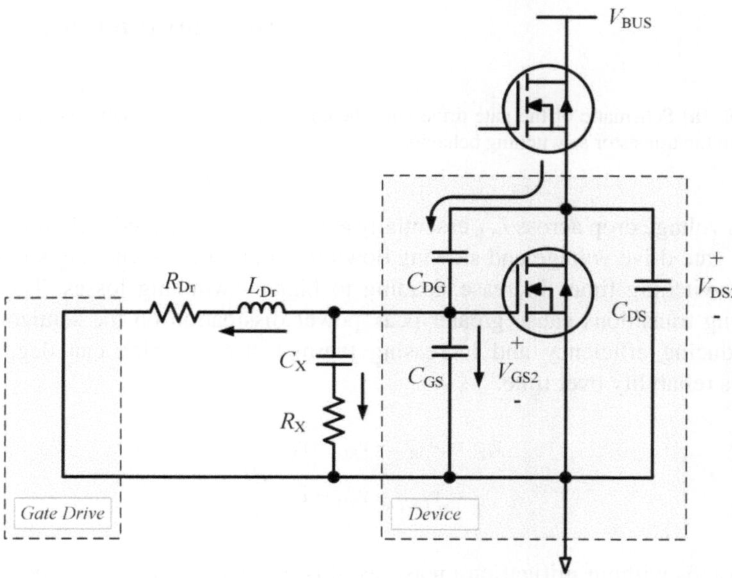

Fig. 4.19 One remedy for the *Miller capacitance* overshoot problem where a series combination of a capacitor C_X and resistor R_X are added in parallel with the gate capacitor C_{GS} to reduce the quality factor of the gate drive signal

transistor switches high currents, particularly under conditions of high *dI/dt*. The resulting voltage drop across L_{CS} introduces a series of negative feedback effects that can disrupt the gate drive signal.

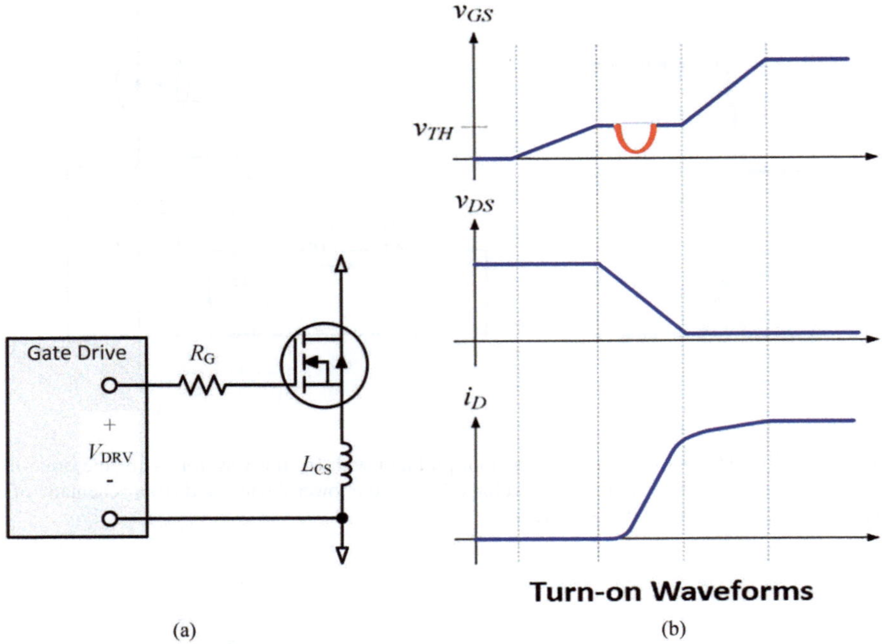

Fig. 4.20 (**a**) Schematic of the gate drive with the common-source inductance (L_{CS}) and (**b**) its impact on the transistor's switching behavior

This voltage drop across L_{CS} essentially acts as a parasitic feedback loop, opposing the gate drive voltage and slowing down the transistor's switching speed. As a result, switching times increase, leading to higher switching losses. The longer switching transitions cause greater peak power dissipation on the semiconductor die, reducing efficiency and increasing thermal stress, which can degrade the device's reliability over time.

$$\begin{aligned} v_{GS} &= v_{DRV} - v_{RG} - v_{LCS} \\ &= v_{DRV} - v_{RG} - L_{CS}\frac{dI_D}{dt} \end{aligned} \tag{4.1}$$

In packages without mitigation measures, this negative feedback can be substantial, resulting in excessive power loss and heat generation during each switching cycle. However, transistors with Kelvin connections, like those shown in Fig. 4.14, significantly reduce this problem. Kelvin connections provide a separate, dedicated path ($L_{SK,WB}$ in Fig. 4.14) for the gate drive signal, bypassing the influence of the source current path (L_{CS}) and minimizing the voltage drop caused by L_{CS}. By isolating the gate control circuit from the high-current source path, these packages reduce switching delays and associated losses, improving overall performance and efficiency.

In summary, common-source inductance is a critical factor in high-current switching applications. Utilizing transistors with Kelvin connections is a highly effective solution to mitigate its adverse effects, resulting in faster switching, lower losses, and enhanced device longevity.

4.4.1 Importance of Managing Temperature Excursions for Device Longevity

Maintaining low-temperature excursions is critical for extending the lifespan of power devices, particularly in demanding applications. This is where soft switching becomes essential, especially for WBG devices like SiC and GaN transistors. Soft switching reduces switching losses and minimizes rapid temperature fluctuations, thereby lowering thermal stress on components.

Thermal excursion mechanisms—such as temperature cycling and thermal shock—accelerate wear and degradation, particularly in solder joints, resistors, capacitors, and other surface-mount assembly (SMA) components. Excessive thermal cycling leads to fatigue, cracking, and eventual failure. To counteract this, thermal management strategies are essential to ensure long-term reliability. Tables 4.5 and 4.6 highlights two key comparisons that illustrate the significance of thermal resistance. First, a comparison between a Si MOSFET and a SiC MOSFET (Table 4.5), and second, a comparison between a Si IGBT and a GaN cascode transistor (Table 4.6).

4.4.1.1 Si MOSFET vs. SiC MOSFET

Both devices are rated for the same current, but the SiC MOSFET's on-resistance (R_{DS_ON}) is just 25% of that of the Si MOSFET. This results in lower conduction losses and greater efficiency. However, the Si MOSFET has nearly 1/10th the thermal resistance ($R_{th,JC}$) of the SiC MOSFET, meaning it can dissipate heat more

Table 4.5 Comparison of the thermal performance for Si and SiC devices

MOSFET	V_{DSS}	I_{D25}	R_{DS_ON}	$R_{th,JC}$	Package
IXFN32N120P (Si)	1200	30	310 mΩ	0.125 C/W	SOT-227B
APT40SM120J (SiC)	1200	32	80 mΩ	1.1 C/W	SOT-227B

Table 4.6 Comparison of the thermal performance for Si and GaN devices

IGBT/cascode	V_{DSS}	I_{D25}	V_{CE_SAT}/R_{DS_ON}	$R_{th,JC}$	Package
IXYA50N65C3 (Si)	650	132	2.0 V	0.25 C/W	TO-263
GS66516B (GaN)	650	20	130 mΩ	1.1 C/W	8×8PQFN

effectively. This comparison highlights the need for superior thermal management when using SiC devices, despite their electrical advantages.

4.4.1.2 GaN Cascode Device vs. Si IGBT

The comparison between a GaN cascode device and a Si IGBT reveals a significant difference in thermal resistance. GaN devices are known for their high switching speeds and low losses, but their higher thermal resistance compared to Si IGBTs requires enhanced cooling solutions. Without adequate thermal management, GaN devices may experience excessive heating, limiting their performance and longevity.

In summary, while WBG devices offer impressive performance gains, their higher thermal resistance presents challenges that must be addressed through effective thermal management. Soft switching plays a pivotal role in reducing temperature excursions, thereby enhancing both device reliability and operational lifespan.

4.4.1.3 Thermal Management Considerations for Power Devices

It is crucial to avoid excessive thermal relief, as it can inadvertently increase the conductive losses of the PCB and reduce the overall lifetime of the unit, particularly through the risk of cold solder joints. Solder joints in surface-mount assembly (SMA) power components, such as resistors and capacitors, inherently have resistance that can contribute to heat buildup. If not properly managed, this can lead to premature failure.

To mitigate these risks, it is important to ensure that resistors and capacitors in the power circuit do not exceed 30% of their nominal power dissipation. By adhering to this guideline, the thermal stress on these components can be minimized, preventing excessive heating and reducing the likelihood of solder joint failure. This approach ensures better thermal stability and extends the longevity of the entire system.

4.5 Conclusion

This chapter examined semiconductor switching devices used in power conversion applications requiring large voltage conversion ratios and high currents exceeding 100 A. In this design space, Si MOSFETs and Si IGBTs are currently employed, depending on application-specific requirements. The chapter also introduced WBG components, which are rapidly becoming more reliable and cost-effective solutions. A significant portion of the text compares three switching devices: (1) Si MOSFET, (2) SiC MOSFET, and (3) SiC Cascode Device. This discussion emphasizes the importance of design engineers carefully reviewing manufacturers' datasheets. For example, while SiC MOSFETs are a highly appealing option for our application, their reliability may be a concern due to their small die size and limited ability to dissipate heat.

Power devices include many non-ideal elements that must be considered. These include parasitic inductances in the transistor leads and the PCB layout. Design engineers must account for these factors to ensure the reliability and performance of their products.

4.6 Supplemental

Some of the images and data contained in this chapter were provided by Havis, Inc. with support from Alec Cook.

Portions of the text and images are from tutorials presented at the *2017 IEEE Energy Conversion Congress and Exposition (ECCE 2017)*, Cincinnati, OH, USA, 2017 and *2022 IEEE Energy Conversion Congress and Exposition (ECCE 2022)*, Detroit, MI, USA, 2022.

This chapter's content is based upon work supported in part by the National Institute for Standards and Technology under Award Number 70NANB22H007 (see U.S. Government's disclaimer in the Acknowledgement section).

The content of this chapter represents the original, creative work of the authors. We utilized OpenAI's ChatGPT to check grammar and spelling and to improve the clarity of certain portions of the text.

References

1. M.J. Scott, K. Zou, J. Wang, C. Chen, M. Su, L. Chen, A gallium nitride switched-capacitor circuit using synchronous rectification. IEEE Trans. Ind. Appl. **49**(3), 1383–1391 (2013). https://doi.org/10.1109/TIA.2013.2255252
2. M.J. Scott, K. Zou, E. Inoa, R. Duarte, Y. Huang, J. Wang, A gallium nitride switched-capacitor power inverter for photovoltaic applications, in *2012 27th Annual IEEE Applied Power Electronics Conference and Exposition (APEC), Orlando, FL, USA*, (2012), pp. 46–52. https://doi.org/10.1109/APEC.2012.6165797
3. M.J. Scott et al., Bidirectional, three-port, three-phase multilevel inverter based on switched-capacitor cells, in *2013 IEEE Energy Conversion Congress and Exposition, Denver, CO, USA*, (2013), pp. 3057–3061. https://doi.org/10.1109/ECCE.2013.6647100
4. C. Li et al., A 2 kW gallium nitride based switched capacitor three-port inverter, in *1st IEEE Workshop on Wide Bandgap Power Devices and Applications, Columbus, OH, USA*, (2013), pp. 119–124. https://doi.org/10.1109/WiPDA.2013.6695577
5. M.J. Scott, J. Li, J. Wang, Applications of gallium nitride in power electronics, in *2013 IEEE Power and Energy Conference at Illinois (PECI), Urbana, IL, USA*, (2013), pp. 1–7. https://doi.org/10.1109/PECI.2013.6506025
6. M.J. Scott et al., Merits of gallium nitride-based power conversion. Semicond. Sci. Technol. **28**(7) (2013). https://doi.org/10.1088/0268-1242/28/7/074013
7. L. Fu, X. Zhang, M. Scott, C. Yao, J. Wang, The evaluation and application of wide bandgap power devices, in *2014 IEEE Conference and Expo Transportation Electrification Asia-Pacific (ITEC Asia-Pacific), Beijing, China*, (2014), pp. 1–5. https://doi.org/10.1109/ITEC-AP.2014.6941195

8. H. Nie, Y. Bai, Q. Cheng, M.J. Scott, Evaluation of conducted EMI in GaN and Si based isolated DC/DC converters, in *2019 IEEE 7th Workshop on Wide Bandgap Power Devices and Applications (WiPDA), Raleigh, NC, USA*, (2019), pp. 251–254. https://doi.org/10.1109/WiPDA46397.2019.8998788
9. W. Perdikakis, M.J. Scott, K.J. Yost, C. Kitzmiller, B. Hall, K.A. Sheets, Comparison of Si and SiC EMI and efficiency in a two-level aerospace motor drive application. IEEE Trans. Transp. Electrif. **6**(4), 1401–1411 (2020). https://doi.org/10.1109/TTE.2020.3010499
10. J. Bendel, *Cascode Configuration Eases Challenge of Applying SiC JFETs* (United Silicon Carbide, Princeton, NJ, 2016). Application Note USCi_AN0004
11. Double Pulse Testing for Power Semiconductor Devices with an Oscilloscope and Arbitrary Function Generator Tektronix
12. B. Sun, Z. Zhang, M.A.E. Andersen, A comparison review of the resonant gate driver in the silicon MOSFET and the GaN transistor application. IEEE Trans. Ind. Appl. **55**(6), 7776 (2019)
13. R. Shahane, S. Belkhode, A. Shukla, Current source gate drivers: Topologies, applications, and future research needs. IEEE Trans. Power Electron. **39**(9), 10731 (2024)
14. H. Peng, H. Peng, Q. Tong, X. Ding, Y. Kang, Review of resonant gate driver from the perspective of driving energy and time. IEEE J. Emerg. Sel. Top. Power Electron. **9**(5), 6344 (2021)
15. B. Sun, Z. Zhang, M.A.E. Andersen, Review of resonant gate driver in power conversion, in *IPEC-Niigata 2018-ECCE Asia Conference Paper*, (2018)
16. A. Isurin, A. Cook, A large MOSFET and IGBT lossless gate driver for high-frequency commutation. https://vanner.com/white-paper/Lossless.pdf
17. Lossless gate driver circuit. US Patent 6,570,416 B1, May 27, 2003

Chapter 5
Design Considerations for Reliable and Cost-Effective Isolation DC–DC Converters

This chapter describes how to control semiconductor switching devices so that power converters meet design objectives. The chapter begins with regulation techniques for power converters, including fixed-frequency and variable-frequency operations. It then reviews several algorithms for implementing pulse-width modulation (PWM) and variable-frequency modulation. Next, the chapter explains how to select the commutation frequency for the chosen modulation strategy. Finally, it examines how regulation is performed, at the system level, in isolated DC–DC converters.

5.1 Design Considerations

An isolation DC–DC converter is a type of power converter that transfers electrical energy between different DC voltage levels while ensuring electrical isolation between its input and output. This isolation is typically achieved using a transformer, which separates the two sides electrically, providing essential protection for sensitive circuits and enhancing overall system safety. Additionally, isolation helps prevent electrical noise from interfering with critical components, making these converters indispensable in applications that require both voltage conversion and high reliability.

Two of the primary benefits of an isolation DC–DC converter are: (1) the separation it provides between the input and output circuits and (2) its ability to change the voltage levels between the input and output circuits. This separation is especially valuable in high-voltage applications, as it limits the risk of high-voltage currents reaching low-voltage control circuits or user interfaces. By design, isolation DC–DC converters protect delicate systems from transient surges, reduce potential interference, and help ensure the safety and longevity of both the equipment and the people who operate it. This unique feature makes them crucial in industries such as

A. Isurin, M. J. Scott, *Practical Design Considerations for Isolated DC-DC Converters*, https://doi.org/10.1007/978-3-031-94427-7_5

automotive, medical, telecommunications, aerospace, and industrial automation, where voltage stability and circuit protection are critical.

In practical terms, isolation DC–DC converters perform both step-up (boost) and step-down (buck) voltage transformations, depending on the application's needs. However, their design often faces constraints, particularly in size, cost, and efficiency. The selection of components, transformer design, control strategy, and thermal management all significantly impact the converter's size, efficiency, and cost. For high-power applications, careful attention to these factors is essential to achieve a design that strikes the optimal balance between performance and affordability.

Designing cost-effective isolation DC–DC converters becomes especially challenging in applications where budget constraints are tight, but isolation and voltage regulation cannot be compromised. Several strategies can help achieve cost-effectiveness in these converters.

First, selecting readily available, standardized components reduces costs and simplifies sourcing and replacement. By choosing lower-cost components that still meet performance standards, designers can make the final product more affordable without sacrificing essential functionality.

In addition to component selection, cost savings can be realized through volume production. Ordering components in bulk or manufacturing converters on a larger scale reduces per-unit costs significantly, as suppliers often offer discounts for high-volume purchases. However, this approach requires careful supplier selection, as comparing pricing, reliability, and warranty options ensures long-term cost benefits while maintaining quality.

Efficiency is another key consideration in designing cost-effective converters. High efficiency (typically between 90% and 98%) minimizes power losses and reduces heat generation, which can decrease the need for extensive cooling systems and thus lower costs. Achieving efficiency within this range strikes the best balance between cost and performance. While targeting near 100% efficiency might seem desirable, it often leads to higher costs due to the more complex materials and designs required. Conversely, a lower efficiency results in greater power losses and heat generation, causing performance issues and added costs for cooling and maintenance. The essential design principle for cost-effective isolation DC–DC converters is to focus on efficient performance within the optimal efficiency range, rather than prioritizing efficiency as the sole goal. The objective is to deliver robust, affordable solutions that meet practical performance standards without unnecessary expense. In high-power applications, this requires a thoughtful approach to thermal management, transformer design, and control methodologies.

For each converter design, a thorough evaluation of the specific application requirements is necessary to ensure that the final design meets both performance and cost objectives. The relationship between cost and efficiency typically follows a curve, where the minimum cost point lies within a practical efficiency range (typically 90–98%). Moving too close to either extreme—whether overly high efficiency or low efficiency—can increase costs due to the additional material and performance demands.

The principles outlined here form the foundation for developing isolation DC–DC converters that meet the needs of both performance and cost efficiency. By leveraging readily available components, considering production volume, selecting efficient suppliers, and targeting an optimal efficiency range, designers can create effective and affordable solutions suitable for a range of high-demand applications.

5.2 Regulation Techniques in Isolated DC–DC Converters: Fixed and Variable Commutation Frequency Methods

The next section compares pulse-width modulation and variable frequency modulation as they apply to isolated DC–DC converters. It talks about the limitations of the switching frequency and other trade-offs. Finally, describes how these modulation strategies are used.

5.2.1 Comprehensive Regulation Methods for Isolated DC–DC Converters: Exploring Fixed and Variable Frequency Techniques

The isolated DC–DC converter is a critical unit designed to provide a stable and regulated output of voltage and current, regardless of varying input and load conditions. Regulation ensures that the converter can maintain precise output parameters, which is vital for system reliability and efficiency. This is especially important in applications involving sensitive electronics, such as aerospace systems, where performance stability is non-negotiable. Effective regulation involves monitoring and adjusting the converter's operation through various control strategies. The goal is to counteract disturbances caused by fluctuations in input voltage, output load, or environmental conditions, such as temperature changes. Poor regulation can lead to issues such as voltage overshoot, output ripple, and instability, which can degrade overall system performance or even cause failure.

In this section, we will examine the fundamental regulation methods commonly employed in isolated DC–DC converters. We'll begin with fixed-commutation frequency techniques, where the switching frequency remains constant, offering predictable performance and simpler filter design. These techniques are often preferred for their straightforward implementation, making them suitable for converters operating within a defined input/output range. We will also touch on other regulation approaches, such as variable-frequency control and pulse-width modulation (PWM), each offering distinct advantages depending on the application's requirements. Furthermore, we'll highlight the trade-offs between efficiency, response time, and EMI (Electromagnetic Interference) when selecting the appropriate regulation method for high-current and high-power designs.

Let's start by diving into fixed-commutation frequency techniques and how they set the foundation for more advanced regulation methods.

5.2.2 *Fixed Commutation Frequency Regulation*

Fixed-commutation frequency regulation offers predictable switching patterns, simplifying filter design and reducing electromagnetic interference (EMI) concerns. Below are the primary techniques used under this category:

5.2.2.1 Pulse-Width Modulation (PWM)

Pulse-width modulation (PWM) is one of the most widely adopted methods for output regulation in isolated DC–DC converters. The technique involves adjusting the duty cycle—the ratio of the on-time to the total switching period—while keeping the switching frequency constant. By varying the pulse width of the control signal, the output voltage or current can be precisely regulated.

- Advantages:
  - High control precision
  - Relatively simple implementation with well-established control ICs
  - Suitable for a wide range of input voltages and output loads
- Challenges:
  - Increased switching losses at higher frequencies
  - Potential EMI issues due to sharp switching edges

5.2.2.2 Phase-Shift Modulation (PSM)

Phase-shift modulation is a variant of PWM, primarily used in full-bridge or half-bridge converter topologies. Rather than adjusting the pulse width directly, the phase difference between gate signals driving opposite legs of the bridge is varied.

- How it works:
  - By shifting the phase between the primary-side switches, the effective duty cycle of the transformer is modulated, controlling the output voltage or current without altering the commutation frequency. This is particularly advantageous in soft-switching applications, where switching losses are minimized.
- Advantages:
  - Reduced switching losses due to zero-voltage switching (ZVS)
  - Lower EMI compared to hard-switching PWM

- Applications:
  - Commonly used in high-power converters, especially where efficiency and thermal performance are critical, such as in aerospace and automotive systems.

5.2.2.3 Combined Phase-Shift and PWM Modulation

This hybrid technique integrates both **phase-shift** and **PWM** control methods, offering enhanced flexibility in managing output parameters. By coordinating the phase-shift between the primary and secondary sides of the transformer with pulse-width adjustments, finer control can be achieved.

- Benefits:
  - Enhanced output regulation under varying load conditions
  - Improved soft-switching range, reducing losses and thermal stress on components
- Use case:
  - Ideal for converters requiring high efficiency across a wide operating range, especially in multi-mode converters where different load conditions demand different control strategies

Each of these techniques offers specific advantages and trade-offs in terms of efficiency, control complexity, and component stress, making the selection of an appropriate method crucial in the design of high-power, high-current isolated DC–DC converters.

5.2.3 Variable Commutation Frequency Regulation

Variable commutation frequency regulation is commonly employed in resonant converter topologies, where efficiency under varying load conditions, especially light loads, is critical. By adjusting the switching frequency in response to load changes, this method reduces switching losses and enhances overall system performance. Figure 5.1 illustrates simplified algorithms for implementing variable frequency regulation. This technique is effective in applications where efficiency under light-load conditions is a priority.

There are three primary approaches to variable frequency regulation:

1. Variable frequency with constant duty cycle
2. Variable frequency with constant on-time or constant off-time pulses
3. Variable frequency with PWM

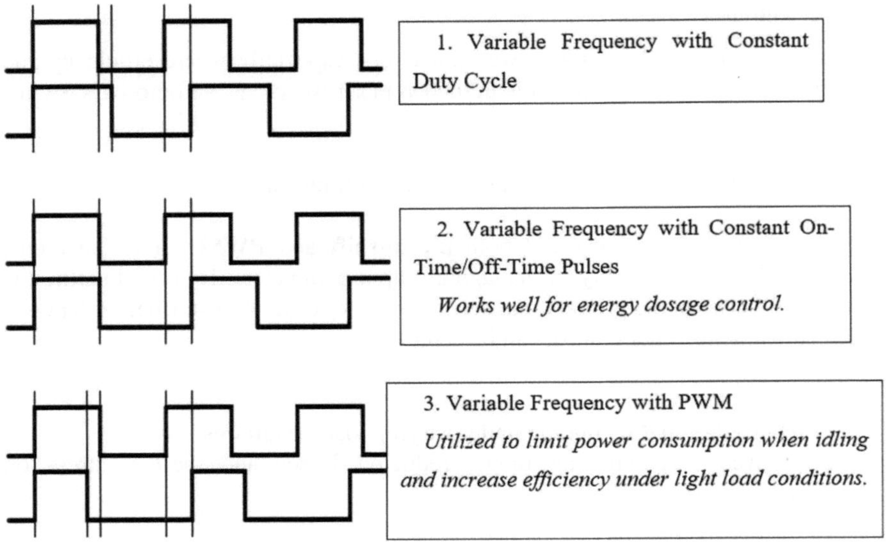

Fig. 5.1 Three simplified algorithms for implementing variable frequency regulation

5.2.3.1 Variable Frequency with Fixed Duty Cycle

In this mode, the **commutation frequency varies**, while the duty cycle remains constant, typically around **48%**, with the remaining **2% reserved for dead time**. This approach is often used in **resonant converters** where the switching frequency aligns with the **natural resonance of the tank circuit**, minimizing switching losses and maximizing energy transfer.

- **Advantages**:
  - High efficiency across a broad load range
  - Simplifies control circuitry by maintaining a constant duty cycle
- **Applications**:
  - **LLC resonant converters** in power supplies for consumer electronics and industrial applications
  - High-current, high-voltage DC–DC converters in automotive systems

5.2.3.2 Fixed Pulse Width with Variable Frequency

In this method, the **pulse width remains constant** while the switching frequency adjusts based on the load. This results in a **variable duty cycle**. It is particularly effective for converters needing **reduced energy consumption** in idle states or **high efficiency at light loads** (typically **5–10%** of nominal load).

- **Advantages**:
  - Significant energy savings during low-power operation
  - Reduced switching losses, enhancing thermal performance
- **Applications**:
  - Suitable for converters that use the **principles of dosed energy transfer**, where small energy packets are delivered during each cycle, ideal for **standby mode** or **battery-powered devices**

5.2.3.3 Hybrid Regulation (Variable Frequency with PWM)

This approach blends **variable frequency control** at light loads (up to **25%** of nominal load) with **PWM** at higher loads. Initially, the frequency decreases as the load drops, maintaining efficiency. Once the frequency reaches a minimum threshold, further regulation is achieved through PWM to prevent frequency saturation or instability.

- **Advantages**:
  - Maximizes efficiency across the entire load range
  - Smooth transitions between light and heavy load conditions, avoiding abrupt shifts in control modes
- **Applications**:
  - Common in multi-mode converters used in renewable energy systems and vehicle electrification, where load conditions vary widely
- **Challenges**:
  - Requires more complex control algorithms and precise timing management to ensure seamless transitions

5.2.3.4 Summary

Variable commutation frequency regulation is an effective strategy for enhancing efficiency, particularly in resonant and soft-switching converters. By tailoring the control approach to the load profile, it enables converters to operate efficiently across diverse conditions, reducing thermal stress and extending the lifespan of components.

5.2.4 *Selecting the Optimal Commutation Frequency*

Determining the optimal commutation frequency in isolated DC–DC converters is a critical design decision, particularly in high-power applications exceeding 2 kW. The frequency impacts several aspects of the converter, including efficiency, power density, component size, and cost. Designers must carefully balance these factors to achieve the desired performance without compromising reliability.

5.2.4.1 Commutation Frequency Constraints and Trade-Offs

There is an interdependence on the switching frequency, the size of the converter via its passive components, and the converter's cost. The design engineer must balance these trade-offs to achieve their design objectives.

- **Passive component size reduction**:
  Higher commutation frequencies allow for smaller capacitors, inductors, and transformers, leading to improved power density and lower material costs. This benefit is especially important in space-constrained applications such as aerospace or electric vehicles.
- **Cost vs. frequency curve**:
  As illustrated in Fig. 5.2, there is a non-linear relationship between frequency and cost. Frequencies above 1 MHz introduce higher costs due to:
  - Advanced filtering requirements to mitigate EMI
  - Increased isolation challenges in high-voltage applications
  - Efficiency losses stemming from switching transitions and parasitic effects, even with soft-switching techniques
- **EMI management**:
  Higher frequencies increase the risk of electromagnetic interference (EMI), necessitating more sophisticated shielding and filtering strategies. Failure to address EMI can lead to regulatory compliance issues and degraded system performance.

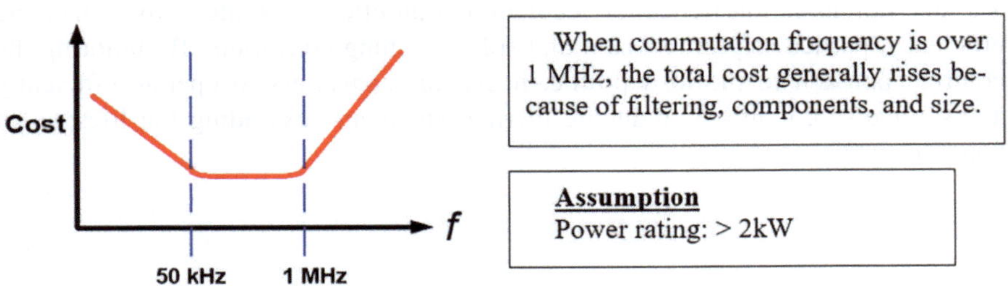

Fig. 5.2 General relationship between cost and frequency for isolated DC–DC converter

- **Practical frequency limits**:
  While soft-switching techniques, such as ZVS or ZCS, can extend the viable frequency range, practical limits exist due to:
  - *Semiconductor losses*: Higher switching frequencies increase turn-on/turn-off losses in transistors, even for advanced devices like GaN or SiC
  - *Thermal management complexity*: Higher frequencies generate more heat, requiring enhanced cooling solutions such as liquid cooling or heat sinks with forced convection

5.2.4.2 Conclusion

Selecting the optimal commutation frequency involves a delicate trade-off between efficiency, cost, power density, and thermal performance. For high-power, high-current isolated DC–DC converters, careful consideration of frequency constraints is essential to balance these factors effectively. The regulation method and frequency must be tailored to the specific application requirements, considering load profile, operational environment, and long-term reliability goals. By understanding and addressing these trade-offs, designers can achieve robust, cost-effective, and high-performing converter solutions tailored to demanding modern applications.

5.2.5 *Fundamental Configurations of Regulation Techniques in Isolated DC–DC Converters*

Isolated DC–DC converters commonly employ a two-stage power conversion process to regulate output parameters while ensuring galvanic isolation between the input and output. This separation enhances safety, prevents ground loops, and enables voltage level adjustments. Let's break down this two-stage process:

1. **Input stage (primary side)**
   The primary side is responsible for converting the input DC voltage into high-frequency AC voltage through switching devices. This stage often incorporates:
   - **EMI filtering**: Since high-frequency switching can generate EMI, filters are often integrated to reduce noise and meet regulatory standards.
2. **Output stage (secondary side)**
   The secondary side rectifies the high-frequency AC voltage back into DC voltage and delivers the final regulated output. Key components include:
   - **Output filtering**: Capacitors and inductors smooth out the rectified voltage, ensuring minimal ripple and stable output

High-Frequency Transformer

Between the input and output stages lies the high-frequency transformer, which serves multiple critical roles:

- **Galvanic isolation**: Protects the load from potential faults on the input side, meeting safety standards in applications like medical devices, aerospace systems, and industrial automation.
- **Voltage transformation**: Steps the voltage up or down, depending on the application's requirements.
- **Size reduction**: The high commutation frequency allows the transformer to be compact, reducing the size of magnetic and capacitive energy storage elements, thereby increasing power density. This compactness is especially beneficial in applications where space and weight are critical, such as electric vehicles and portable power systems.

Summary

By understanding the two-stage conversion process and the role of high-frequency transformers, designers can select and optimize regulation techniques that balance efficiency, cost, size, and performance.

5.2.5.1 Primary-Side Regulation

Figure 5.3 shows a block diagram for primary-side regulation. Under this type of regulation, control components are positioned in the input section between the DC source and the high-frequency transformer. This approach is prevalent in topologies such as:

- LLC resonant converters
- Phase-shifted full-bridge (PSFB) converters
- PWM-based resonant converters

Advantages of Primary-Side Regulation

1. Excellent control over output parameters:
   Despite the transformer providing galvanic isolation, the primary-side regulation effectively manages output voltage and current through precise control of the switching elements.

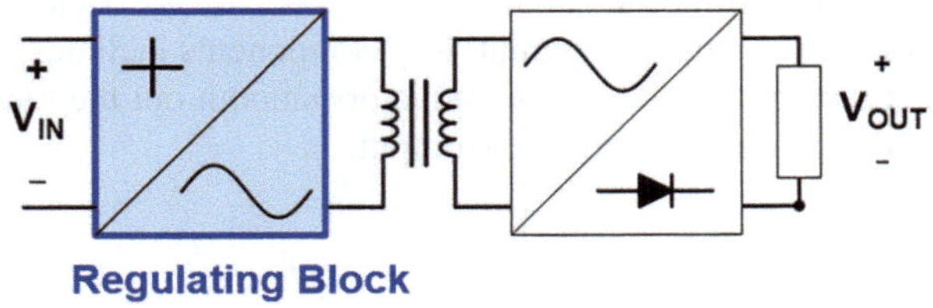

Fig. 5.3 Two-stage isolated DC–DC converter with primary-side regulation

2. Simplicity:
   The regulation circuit remains on the primary side, reducing complexity on the secondary side. This simplicity leads to:
   - Easier feedback loop design
   - Reduced component count on the secondary side, lowering the risk of system failures
3. High power density:
   By employing high commutation frequencies, designers can significantly reduce the physical size of magnetic components, particularly:
   - Transformers: Smaller core sizes due to reduced flux swings
   - Inductors: With lower inductance values, resulting in smaller winding sizes
4. Cost efficiency:
   The reduction in magnetic and passive component sizes translates to lower material costs, making this method ideal for high-power applications exceeding 2 kW, such as:
   - Industrial power supplies
   - Renewable energy systems
   - Electric vehicle chargers

Considerations and Trade-Offs

1. EMI challenges: High-frequency switching can lead to increased electromagnetic interference (EMI), necessitating effective filtering solutions.
2. Limited secondary control: Since regulation is entirely on the primary side, managing secondary-side dynamics can be more challenging, especially under rapidly changing load conditions or in systems requiring high accuracy.
3. Feedback isolation: Optocouplers or magnetic coupling are often necessary to transfer feedback signals from the secondary side to the primary controller, introducing minor latency or design complexity.

Application Examples

Primary-side regulation is particularly beneficial in applications requiring:

- High power efficiency with minimal output ripple
- Compact designs with stringent size and weight constraints, such as aerospace and portable power systems

Conclusion

Primary-side regulation offers a robust, cost-effective, and efficient method for controlling isolated DC–DC converters in high-power applications. Its simplicity and ability to leverage high frequencies for compact designs make it a popular choice in modern power electronics systems.

5.2.5.2 Secondary-Side Regulation

Figure 5.4 is the system diagram for secondary-side regulation. For this method, control is applied to the output stage, where switching devices regulate the output voltage and current. This approach is frequently used in applications with:

- Low input voltage and high input current scenarios
- Step-up (boost) applications require significant voltage transformation from the input to the output

Advantages of Secondary-Side Regulation

1. Efficient for high current, low voltage inputs:
   - Ideal for converters where the input voltage is low and the current is high, such as in battery-powered systems and fuel cells.
   - Reduces primary-side stress by shifting control and switching losses to the output stage.
2. Precise output control:
   - Direct regulation on the output stage provides finer control over output voltage and current, especially under varying load conditions.
   - Suitable for applications requiring tight voltage tolerances, such as data center power supplies and telecommunication equipment.
3. Step-up applications:
   - Well-suited for scenarios where significant voltage boosts are needed, such as in solar inverters or electric vehicle chargers.

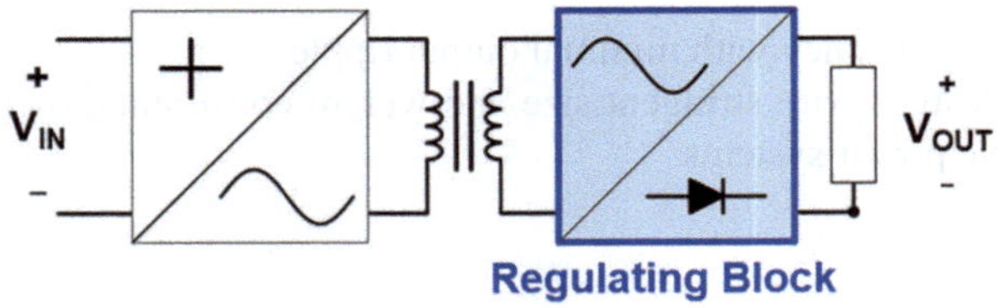

Fig. 5.4 Two-stage isolated DC–DC converter with secondary-side regulation

Challenges and Trade-Offs

1. Larger transformer design:
   - Since the primary-side voltage directly corresponds to the input voltage, the transformer's flux density is highly dependent on input variations.
   - To accommodate higher flux density, transformers need to be larger and more robust, increasing their cost and weight.
2. Complex transformer design:
   - Handling wide input voltage ranges requires careful transformer design to avoid core saturation and excessive losses.
   - This complexity can increase design time and may necessitate advanced magnetic materials or custom core shapes.
3. Increased EMI sensitivity:
   - High current on the primary side can introduce electromagnetic interference (EMI) challenges, requiring shielding and filtering solutions.

Applications

- Step-up converters: Frequently used in boost-type converters, where the output voltage needs to be significantly higher than the input
- High-current systems: Common in battery chargers, renewable energy systems, and high-power LED drivers, where the input voltage is typically low
- Low-voltage DC bus systems: Ideal for systems operating with a 48 V input but requiring high-voltage outputs for distribution or storage

Conclusion

Secondary-side regulation is an effective strategy for low-voltage, high-current applications and step-up converters where precise output control is essential. Despite the increased complexity in transformer design, it offers significant benefits in terms of output precision and efficiency in demanding applications. However, designers must carefully consider transformer size, flux density management, and EMI mitigation to ensure optimal performance.

5.2.5.3 Dual-Side Regulation

The system diagram for dual-side regulation appears in Fig. 5.5. This control method is applied to both the primary and secondary sides of the converter, allowing for:

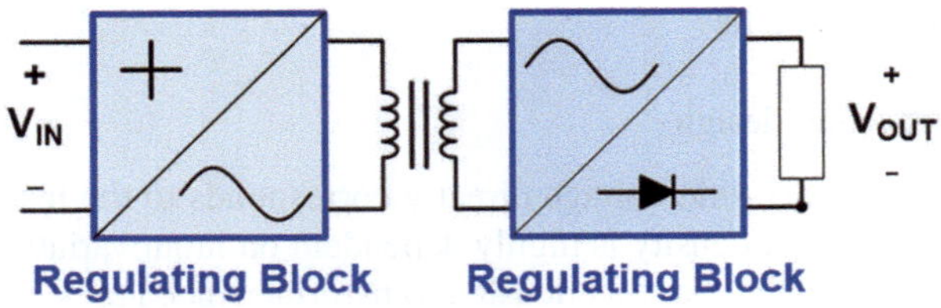

Fig. 5.5 Two-stage isolated DC–DC converter with dual-side regulation

- Precise management of input and output voltages
- Enhanced power flow control over wide input-to-output voltage ranges

This configuration is especially useful for applications requiring:

- Broad voltage requirements
- Bidirectional power flow, such as in battery energy storage systems and automotive powertrains

Advantages of Dual-Side Regulation

1. Wide voltage range handling:
   - Ideal for applications where input and output voltages vary significantly, such as renewable energy systems (e.g., solar inverters and wind turbines) and electric vehicle powertrains.
   - Supports seamless operation across varying voltage levels, making it suitable for hybrid grid-connected and off-grid systems.
2. Bidirectional power flow:
   - Particularly beneficial for energy storage systems and vehicle-to-grid (V2G) applications, where power must flow efficiently in both directions.
   - This is critical for regenerative braking in electric vehicles and dynamic energy management in smart grids.
3. Precise control of power flow:
   - Dual-side regulation enables fine-tuned power management, allowing converters to dynamically adjust to varying load and supply conditions.
   - This precise control improves overall system efficiency and response times under fluctuating loads.

Common Topologies

1. CLLC resonant converters:
   - Frequently used in high-efficiency and high-frequency applications requiring zero-voltage switching (ZVS) or zero-current switching (ZCS)
2. Dual current-fed converters:
   - Suitable for low-voltage, high-current systems, offering robust control on both sides for better efficiency
3. Dual active bridge (DAB) converters:
   - Suitable high-power bidirectional systems
4. Dual active half-bridge (DAHB) converters:
   - Used in applications with minimum size and weight

Design Considerations

1. Complex control strategy:
   - Implementing dual-side regulation requires a more sophisticated control algorithm, increasing design complexity.
   - Precise coordination between the primary and secondary sides is essential to avoid control loop instability.
2. Increased component count:
   - More components, such as sensors, controllers, and switching devices, are required, leading to higher costs and more intricate PCB layouts.
3. Thermal management:
   - With both sides regulated, thermal dissipation must be carefully managed, especially in high-power applications, to avoid overheating and ensure long-term reliability.

Applications

- Battery energy storage systems: Enables efficient charging and discharging under varying grid conditions
- Electric vehicle powertrains: Supports bidirectional power transfer for features like regenerative braking and V2G capabilities
- Renewable energy systems: Ensures seamless energy transfer between sources and storage units, even with fluctuating input voltages

Summary

Dual-side regulation offers unmatched flexibility, precision, and efficiency, making it ideal for converters requiring bidirectional energy flow or wide input-to-output voltage ranges. Despite its complexity, it is a powerful solution for advanced power systems, enabling dynamic adaptation to diverse operating conditions while maximizing system performance.

5.3 Conclusion

This chapter reviewed regulation techniques for isolated DC–DC converters. First, it offered recommendations to make hardware more reliable and cost-effective. The second portion of the chapter considered regulation techniques. This included fixed frequency and variable frequency methods. Finally, it discussed the primary-side, secondary-side, and dual-side regulation system. It described the applications of each approach, described their advantages, and discussed the challenges and trade-offs to consider. This section serves as a foundation for Chap. 6, which will utilize these concepts as it evaluates six types of isolated DC–DC converters.

5.4 Supplemental

The content of this chapter represents the original, creative work of the authors. We utilized OpenAI's ChatGPT to check grammar and spelling and to improve the clarity of certain portions of the text. Portions of the text and images are from tutorials that we presented at the *2017 IEEE Energy Conversion Congress and Exposition (ECCE 2017)*, Cincinnati, OH, USA, 2017 and *2022 IEEE Energy Conversion Congress and Exposition (ECCE 2022)*, Detroit, MI, USA, 2022.

Havis, Inc., with support from Alec Cook, provided some of the images and data in this chapter.

This chapter's content is based upon work supported in part by the National Institute for Standards and Technology under Award Number 70NANB22H007 (see U.S. Government's disclaimer in the Acknowledgement section).

Chapter 6
Isolated DC–DC Converter Topologies

This chapter presents the design criteria we created to compare power converters for a given application. Our criteria are based on the characteristics of an 'ideal' power converter, and they highlight each topology's disadvantages and advantages.

We apply our design criteria to six fundamental isolated DC–DC converter topologies that are suitable for applications where the output power exceeds 2–3 kW and low-voltage side currents are above 100–200 A. They are: (1) Resonant LLC Converter; (2) Phase-Shift (PS) Converter, (3) Dual Active Bridge (DAB) Converter; (4) Resonant with Pulse Width Modulation (R-PWM) Converter; (5) Dual Current-Fed with Active Clamp (DCAC) Converter; and (6) Dual Current-Fed with Switched Capacitors (DCSC) Converter.

We discuss these topologies in order of their development, starting with the oldest and concluding with the most recent. These topologies are evaluated based on electrical, operational, and application-specific parameters. The resulting information enables a comparative analysis of these topologies, guiding design engineers in selecting the most appropriate configuration. This evaluation approach also offers a framework for preliminary estimation, helping engineers develop a well-informed converter design.

6.1 Definitions of Comparison

Isolated DC–DC converters are used in a variety of applications that range in power levels from milliwatts to megawatts. Designers have many 'basic' topologies available to use in their applications, and each topology has its own set of disadvantages and advantages. A survey of patents and literature reveals numerous variations and modifications to these basic topologies to tailor their performance for a specific design target. This is because no basic topology is perfect and rarely is it suitable, without alteration, for an engineer's design objective. Finally, the design engineer

A. Isurin, M. J. Scott, *Practical Design Considerations for Isolated DC-DC Converters*, https://doi.org/10.1007/978-3-031-94427-7_6

must select a topology to use, based on its advantages and disadvantages, with a minimum number of modifications, to satisfy design criteria. In this chapter, we offer a methodology to select an isolated DC–DC converter for a given application.

There are numerous parameters to consider when selecting an isolated DC–DC converter. However, we will focus on the key factors that align with the engineering philosophy that **the primary requirements for any power converter are: (1) cost; (2) reliability; and (3) compatibility**. While the topology alone does not dictate the final cost, it is essential to consider the entire system, including filtering, to get a complete picture. We mean more than the material and engineering costs of a power stage. We want to examine the whole picture of the system. This must include the electromagnetic interference (EMI) filter because a power stage will not operate without an EMI filter in a practical application. This filter can be very expensive, impact cost, and lower power density to meet regulatory standards, such as FCC Part 15 and CISPR 25 for commercial applications and MIL-STD-461, and DO-160 for military products.

Our book focuses on converters with power levels of 2 kW or more and currents in the hundreds of amps. These topologies belong to a family of soft-switching converters, to varying degrees. We will examine their behavior in the context of implementing an isolated DC–DC converter that meets the typical requirements for high-current isolated converters in the transportation sector. Currently, the transportation industry primarily demands the technical specifications shown in Table 6.1 with a high-voltage side ranging from 270 to 850 VDC and a low-voltage side from 8 to 60 VDC, with current capabilities of 200–800 ADC. Additionally, it is preferable for these converters to operate in bidirectional mode.

For our comparison of these topologies, we will use emerging specifications that represent the average requirements for isolated DC–DC converters in the transportation sector. We will focus solely on unidirectional operation, either in step-down or step-up modes. In step-down mode, the input voltage will range from 550 to 800 VDC, with an output of 30 VDC at 400 A, as shown in Table 6.2. In step-up mode, the input voltage will range from 20 to 30 VDC with a maximum current of 400 A, stepping up to an output voltage of 700 VDC, as shown in Table 6.3.

These converters must:

- Operate effectively from 10% load to their maximum load and be capable of changing modes (e.g., from voltage source mode to current source mode) during light load or overload conditions. It is highly desirable for the converter to also function in idle mode.

Table 6.1 General electrical requirements for DC–DC converters in the transportation sector

High-voltage side	270–850 VDC
Low-voltage side	8–60 VDC
Low-voltage side	200–800 ADC
Standards	CISPR 25
Mode	Electrical isolation Bidirectional operation (preferred)

Table 6.2 Parameters for the step-down evaluation of the converters

Input side	550–800 VDC
Output side	30 V at 400 ADC
Standards	CISPR 25

Table 6.3 Parameters for the step-up evaluation of the converters

Input side	20–30 VDC at 400 ADC
Output side	700 VDC
Standards	CISPR 25

- Have a minimum lifetime of 10 years.
- Perform reliably when the load is suddenly disconnected while operating at maximum power (i.e., in load dump mode).
- Meet or exceed Class 4 standards of CISPR25.

For our comparison, we used the simulation software NL5, assuming that all converters have the same efficiency. We recognize that simulations do not fully capture the real-life behavior of converters and only reflect the basic principles of real-world components. Furthermore, the efficiency calculations do not account for conductive losses in the busbars, printed circuit boards (PCBs), or power connections.

We begin by examining basic topologies, defining their parameters, listing their main advantages and disadvantages, and describing their application areas. If a basic topology does not meet our requirements, we will explore the possibility of adapting it. All topologies use the same semiconductors on the high-voltage side and employ synchronous rectification on the low-voltage side.

Our book does not explore topologies that require two or more transformers or parallel-connected power stages, as they are prohibitively expensive. We also exclude multi-level topologies, which are not cost-effective below 1000 VDC. A design engineer would need a compelling justification to choose a multi-level topology below this voltage level, as it increases cost, reduces reliability, and complicates control. Additionally, the authors are confident that cost-effective isolated DC–DC converters with power levels below 1 kW can be implemented using a power stage with just one or two active switches, not including synchronous rectification.

Our philosophy—the main requirements for power converters are cost, reliability, and compatibility—drives the analysis in this book. We extend this philosophy to say that engineering is a tool that businesses use to make a profit. This means that the design engineer must always consider the cost of their design. It is the main point of consideration for any practical power converter. For the transportation and industrial areas, cost is the number one concern while it is number two for military and aviation applications, behind weight. Therefore, it does not make sense to compare topologies without considering their costs. **To produce a cost-effective design is one of the arts of engineering.** It depends on the knowledge and the skill level of the design engineer, even more so when the application requires high current. Therefore, **cost is a higher priority point for comparison**. We are going to make an estimation of only the power parts of the power converter, the power stage itself, filters, auxiliary circuits, and how these elements influence each other. The cost of all other very important parts, the busbars, PCB, cooling system, connections, control board, housekeeping power supply, and many other important parts are very individual, very strong influence of engineering art, and are not considered in this work.

A major challenge of this requirement, cost-effective, is the magnetic components required for high-current operation. Along with designing the magnetic elements, the engineer must create high-current connections with low loss and mechanical supports for the magnetics, both of which require customization and creativity. The size and quantity of these components are not trivial, and they can significantly influence the cost of the topology. To define this influence, we created the **Coefficient on big size of parts.** It shows how big components influence the cost of the converter. This coefficient and power transformer were estimated by our experience. For the components, we used the prices listed from an online vendor's website and assumed that we needed enough parts for 1000 units per year. For each topology, we will use the same parts. Our target with this analysis is to provide another metric to help the designer to compare topologies.

Of course, we do not claim to have selected the best components, but the analysis is sufficiently thorough to illustrate our approach. It can be adapted by design engineers to evaluate hardware for their specific applications.

Isolated DC–DC converters are power converters that move energy from a power source to a load. This process is implemented by a switch-mode power supply (SMPS), and an artifact of an SMPS is that the converter interrupts current during each switching cycle. It is important to know the amount of time (t_{FLOW}) that the energy moves from the source to the load relative to the switching period (T). We created another parameter referred to as the **duty cycle of power conversion (*P–C*)**:

$$P - C = \frac{t_{FLOW}}{T} \cdot 100\%. \tag{6.1}$$

This is a very important parameter for our comparison because a converter becomes more efficient as this percentage increases. The ideal case is when *P–C* equals 100%.

The power stage of an isolated DC–DC converter includes semiconductor power devices, an isolation power transformer, and passive components. The power transformer is a very important part of the power stage. It is relatively the most expensive part by itself and this does not consider the mechanical support for the transformer and the high-current connections to the circuitry. As of 2025, this cost is around 20–35% for all of the parts exclusively in the power stage. It is worth mentioning that our analysis in 2017 estimated these costs to be roughly 15–20% of the power stage. Therefore, we need to know how to use the power transformer efficiently to minimize cost. For this target, we created another parameter that correlates the output power (P_{OUT}) to the volt-amp (VA) rating of the transformer the **Coefficient of Power Transformer Use**:

$$Co\text{-}XMFR = \frac{P_{\text{OUT}}}{V \cdot A} \cdot 100\%. \tag{6.2}$$

This is a very important parameter for our comparison because the converter becomes more efficient as this percentage increases. The ideal case is when *Co-XFMR* is 100%.

The power converter is a system that demands control and regulation of the output parameters, namely the voltage and current. This regulation must be fast and stable, and it must minimize the changes to the output parameters of the converter relative to big changes in the outside parameters of the converter—changes to the level of load or the input voltage. We need to know how the converter responds to these changes. The converter regulates the output parameters by changing the duty cycle of the pulse width modulation (PWM), changing commutation frequency, or changing both at the same time. Therefore, we use another parameter in our evaluation—the percentage of change in the converter's PWM duty cycle that results in a 1% change in the output voltage. We refer to this as the stability **coefficient-Stab/Coeff**. When this parameter is high, then it is relatively easy to provide good stability in the control loop.

Estimation of the **level of EMI** of a topology is a point for comparison. However, we will not do an in-depth treatment of this subject in the book, but we will discuss how the requirements for EMI influence the cost of the converter. EMI mitigation must be considered during the design phase of the converter. If not, it will waste time and money during the prototyping phase.

The possibility or quantity of modification of a topology is a factor for comparison. The design engineer makes modifications or develops new topologies when existing topologies do not reflect the specifications that the design engineer needs to implement. Obviously, we do not have ideal power converters at the present time, but when a topology has many numbers of hardware modifications, it has to draw the attention of the design engineer. In other words, the topology with fewer modifications is preferable relative to the topology which has numerous modifications. However, modifications to the control algorithms and software do not influence the production costs, but hardware modifications can significantly change the production costs and most of the time, these costs go up. Simple estimation of the reliability of the converter can be done by the **reaction of the converter to an overload without compensation by the control**.

6.2 Resonant LLC Converter

The LLC converter topology belongs to the family of resonant topologies that use variable frequency control. This type of power converter started being used in the 1970s. You can read more about the history and detailed explanation of this topology in [1]. Figure 6.1 shows the basic schematic of this topology. Generally, the Resonant LLC Converter is a unidirectional power converter, but theoretically, most power converters can operate in reverse. As a result, it is possible to use this topology for bidirectional power conversion [2, 3]. One of the key advantages of the LLC topology is that it achieves soft switching across the entire range of load conditions and input voltages. The LLC topology can be implemented using a full-bridge or half-bridge configuration (the half-bridge configuration appears in Fig. 6.1).

Figure 6.2 shows a Resonant LLC Converter with a full-bridge regulating section. The choice between these two configurations depends on the input voltage and output power requirements. It is worth noting that the half-bridge configuration is generally preferred due to its lower cost. This topology has been widely studied, with many publications discussing its description, optimization, modifications [4, 5], and the calculation of different variations [6–9]. Given the extensive amount of literature on this converter, we will only briefly review the LLC Converter with half-bridge regulation and recommend the reader consider the listed references for additional information.

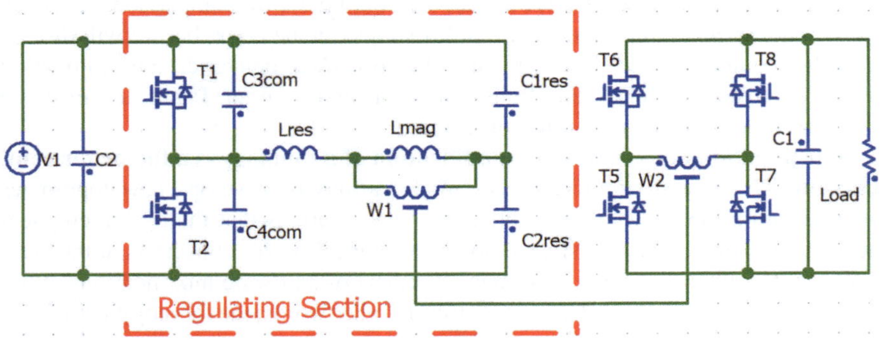

Fig. 6.1 The schematic of a basic Resonant LLC Converter with a half-bridge regulating section

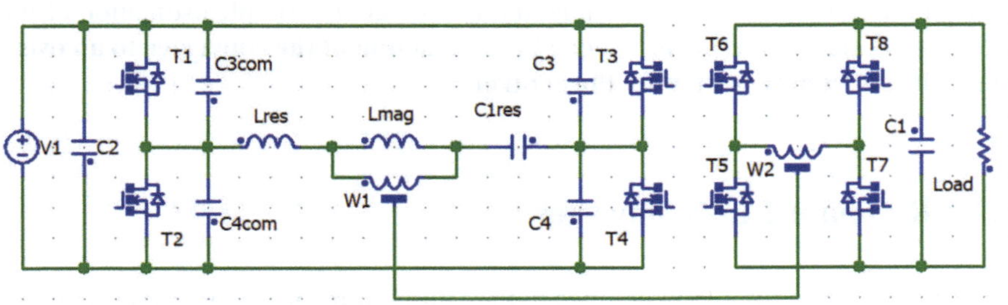

Fig. 6.2 The Resonant LLC Converter with a full-bridge regulating section

6.2.1 Operation of the LLC Topology

Figure 6.3 shows the simulation waveforms of the converter in Fig. 6.1 when the commutation frequency is lower than the resonant frequency, resulting in the maximum output for this simulation setup. Figure 6.4a–d illustrate the equivalent circuits of each operation mode during one half-cycle of operation.

Starting with Fig. 6.4a, which shows the time segment between t0 and t1, the device T2 turns on at t0 under pure ZVS conditions and begins to conduct in synchronous rectification mode. During this period, the energy of the magnetizing

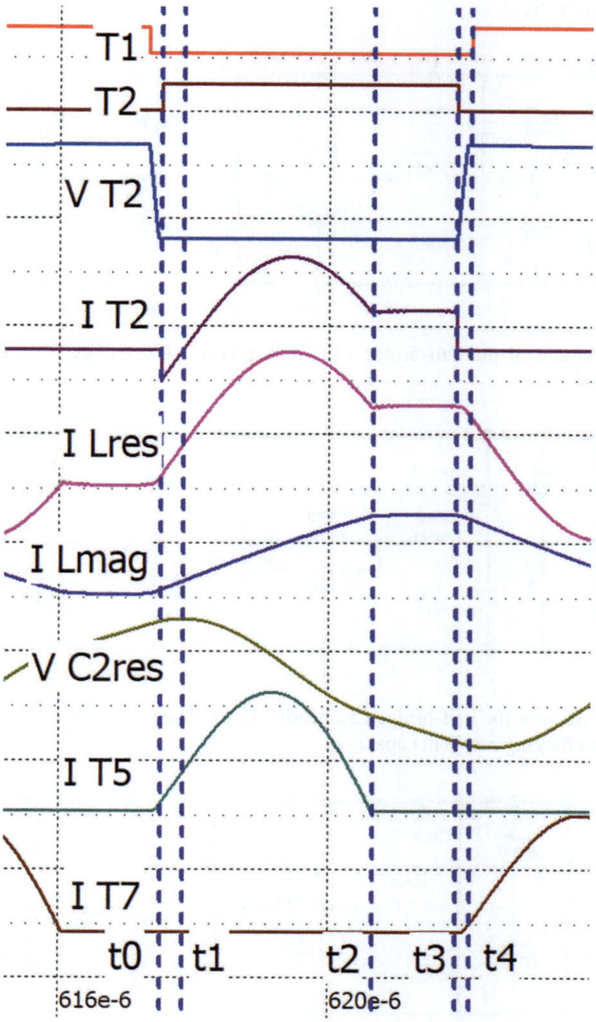

Fig. 6.3 The simulation waveforms of the Resonant LLC Converter operating at a switching frequency below its resonant frequency

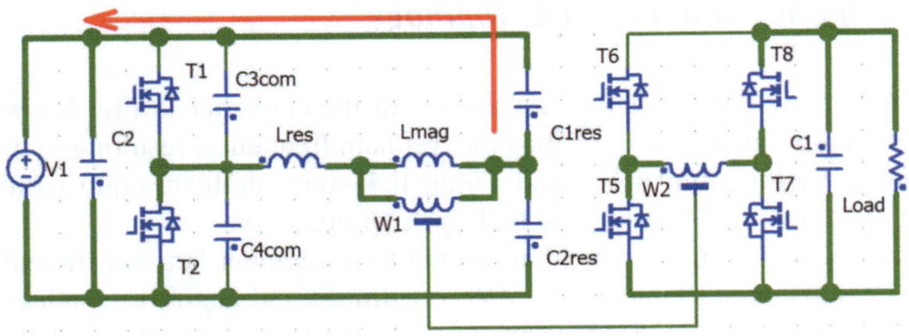

Fig. 6.4 (**a**) Equivalent circuit of the half-bridge LLC during t0–t1 ZVS Turn-on of T2 and continued charging of C2res2

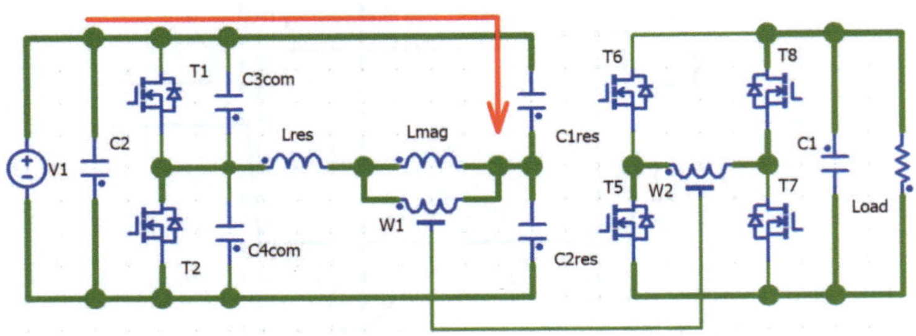

(**b**) Equivalent circuit of the half-bridge LLC during t1–t2. The energy is transferred from the source to the load

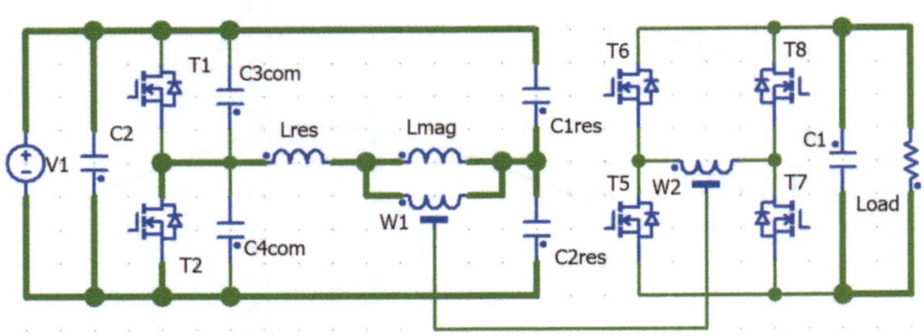

(**c**) Equivalent circuit of the half-bridge LLC during t2–t3. The end of the energy transfer process and charging/discharging resonant capacitors

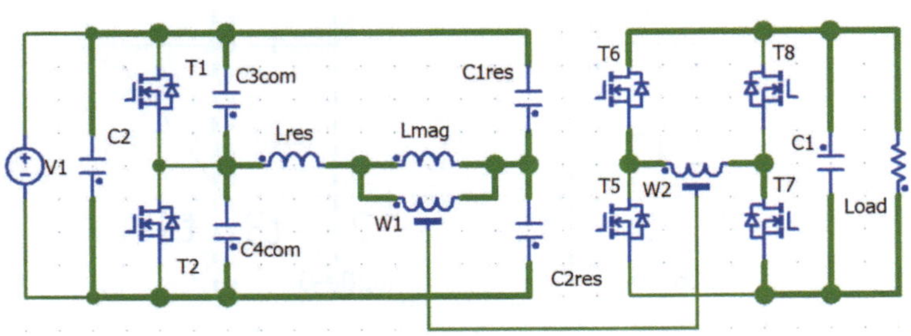

(**d**) Equivalent circuit of the half-bridge LLC during t3–t4. T2 turns off under ZVS conditions, and T1 turns on under ZVS conditions, and the converter starts a new half-cycle of power conversion

inductor (Lmag) continues to charge the resonant capacitor C2res. At time t1, the voltage across this capacitor C2res reaches its maximum value. Part of this energy returns to the power source (V1, C2), while another portion is transferred through the power transformer, passing through T5 and T8 to the load.

At time t1, energy begins to transfer from the source to the load and continues until the time t2 as shown in Fig. 6.4b. During this interval, the magnetizing inductor continues to draw energy from the source. At time t2, the energy transfer process from the source to the load stops. During the period between t2 and t3, the energy from the source is used for charging and discharging the resonant capacitor, as illustrated in Fig. 6.4c.

At time t3, device T2 turns off, interrupting the magnetizing current, which has a relatively high value at this point. This event occurs under pure ZVS conditions and is facilitated by the commutation capacitors C3com and C4com, as shown in Fig. 6.4d. At time t4, the device T1 turns on under ZVS conditions, and a new half-cycle of power conversion begins. This new half-cycle is a mirror image of the half-cycle occurring between t0 and t4.

6.2.2 Practical Issues and Calculation of Main Components

Now that we understand the basic operation of the LLC converter, let's focus on some important practical issues. As the output power increases, the value of the resonant capacitor increases, while the value of the resonant inductor decreases if we aim to maintain the same commutation frequency. Although increasing the commutation frequency is an option, certain parameters primarily depend on the output power and cannot be changed significantly without incurring higher costs. These parameters include the leakage inductance of the power transformer and the inductance of the output circuits. While we have some flexibility in modifying the transformer's leakage inductance, but it is more challenging to change the output circuit's inductance, especially when dealing with high output currents of 200 A or more. This limitation is constrained by the physical size of the components and their connections. Let's consider these conditions for our target output: 30 V and 400 A. The inductance of this circuit, in practical implementation, typically ranges between 5 and 20 nH. An inductance higher than 20 nH is generally not desirable for effective implementation. For this reason, we only consider surface mount components. For our analysis, we'll use an average value of 12 nH and have the input voltage vary between 550 and 800 V.

Next, let's perform a basic calculation of the resonant circuit and the transformer ratio, following the methodology described in [6]. We won't be optimizing the converter's design but rather, we will demonstrate its basic principles. The ratio (m) between $L_P = L_{mag} + L_{res}$ and L_{res} is chosen as 6.

$$m = \frac{L_P}{L_{res}} = \frac{L_{mag} + L_{res}}{L_{res}} \tag{6.3}$$

The minimum gain ($M_{\min}$) is calculated as:

$$M\min = \sqrt{\frac{m}{m-1}} = \sqrt{\frac{6}{6-1}} = 1.095 \tag{6.4}$$

The turns-ratio (N) of the power transformer is:

$$N = \frac{\text{Vin}\max}{2\text{Vout}} \times M\min = \frac{800}{2x30} \cdot 1.095 = 14.6\,\text{turns}. \tag{6.5}$$

We chose 15 as the turns-ratio, considering we have 1 turn on the low-voltage side of the transformer.

Using the parameters in Table 6.2, the equivalent load resistance (R_{AC}) is:

$$R_{AC} = \frac{8N^2V_{\text{OUT}}^2}{\pi^2 P_{\text{OUT}}} = \frac{8 \bullet 15^2 \bullet 30^2}{3.141^2 \bullet 12{,}000} = 13.69\,\Omega \tag{6.6}$$

Given that the commutation frequency is 130 kHz and the quality factor (Q) is 0.4 [6], the resonant capacitor $C_{1\text{RES}}$ is calculated as:

$$C_{1\text{RES}} = 1/2\pi Qf\,\text{R}_{\text{ac}} = 10^6/2 \times 3.14 \times 0.4 \times 130 \times 10^3 \times 13.69 = 0.224\,\mu\text{F}. \tag{6.7}$$

The resonant inductor L_{RES} is:

$$L_{\text{RES}} = 1/(2\pi f)^2\,\text{C}_{1\text{RES}} = 10^6/\left(39.44 \times 225 \times 10^8 \times 0.22 \times 10^{-6}\right) = 6.8\ \mu\text{H}. \tag{6.8}$$

For $m = 6$: $L_P = 6{\cdot}6.8 = 40.8\ \mu\text{H}$ and $L_{\text{mag}} = 40.8–6.8 = 34\ \mu\text{H}$.

Figure 6.5 is the equivalent AC circuit of the Resonant LLC Converter. The total leakage inductance of the power transformer is 2.2 μH without taps and connections. Figure 6.6 shows the AC equivalent circuit of the real Resonant LLC Converter. Lres has been separated into the transformer's leakage inductance, L_{leak}, and an additional inductance, L_{add}. L1 on the secondary is added to model the stray inductance of the PCB. The new ratio m is approximately 9.3. This change significantly affects the converter's behavior, so, design engineers must carefully consider this factor during the development phase to avoid issues when the prototype is ready.

6.2.3 Simulation Results Analysis

Now, we can consider the results of simulations under different operating conditions, with varying input voltages and load levels. This topology operates with complete soft switching, so for calculating losses, **we only considered the conductive**

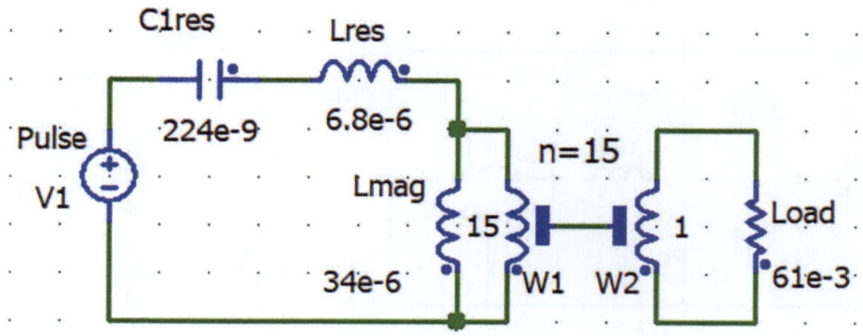

Fig. 6.5 The AC equivalent circuit of the real LLC resonant converter

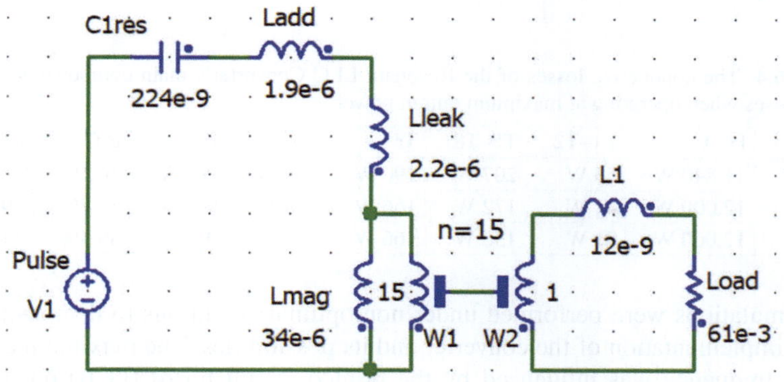

Fig. 6.6 The AC equivalent circuit of the real LLC resonant converter, where the ratio *m* is approximately 9.3

losses of the main power components. Figure 6.7 shows the schematic used for the simulation and calculations under the following conditions:

- **Input voltage (V1)**: 550–800 V
- **Output voltage (VLoad)**: 30 V
- **Maximum power**: 12 kW
- **Resonance frequency**: 131 kHz
- **Resonance inductor (Lres)**: 4 µH
- **Magnetizing inductor (Lmag)**: 32 µH
- **Stray inductance in the PCB (L_1, L_2)**: 12 nH each
- **Output capacitor (C1)**: 1000 µF
- **Resonance capacitors (C1res and C2res)**: 110 nF each
- **Commutation capacitors (C3com and C4com)**: 2.5 nF each
- **Power switches T1 and T2**: Two UF4SC120030K4S transistors connected in parallel
- **Power switches T5–T8**: 10 IAUC120N06S5N032ATMA1 transistors connected in parallel
- **Turns ratio of power transformer**: 15

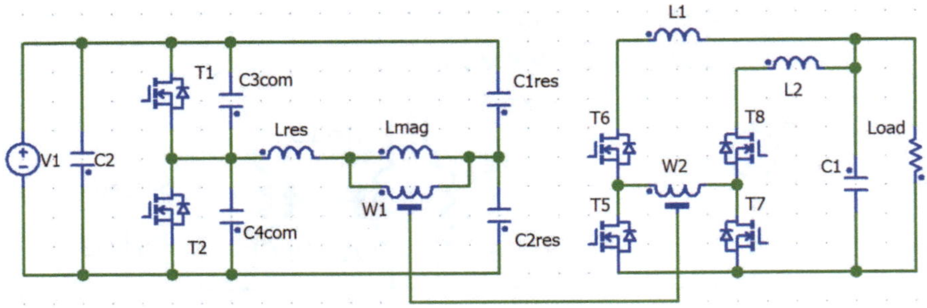

Fig. 6.7 The schematic of the Resonant LLC Converter used for the simulation and calculations in our analysis

Table 6.4 The conductive losses of the Resonant LLC Converter's main components and their total losses when operating at maximum output power

Vin	Pout	T1–T2	T5–T8	Tr	Total	Eff	P-C	f_{COM}
550 V	11,840 W	55 W	207 W	190 W	452 W	96.3%	56.5%	81 kHz
650 V	12,000 W	41 W	172 W	166 W	379 W	96.9%	68.7%	91 kHz
800 V	12,000 W	31 W	136 W	166 W	333 W	97.3%	89.3%	119 kHz

All simulations were performed under non-optimal conditions to demonstrate the basic implementation of the converter and its possibilities. The maximum commutation frequency was influenced by the printed circuit board (PCB) on the low-voltage side and the additional losses in connections of the power transformer. For this current level, maintaining a frequency over 100 kHz is challenging and increases costs. Efficiency decreases by 1% or more when the commutation frequency is 150 kHz or higher.

The simulations examine various conditions with different input voltages and load levels. Since this topology achieves complete soft switching, we only calculated the conductive losses of the main power components [9]. In a real-life implementation, it would be necessary to include conductive losses from the PCB, busbars, and filters, which could contribute an additional 1% or more, depending on cost requirements and the effectiveness of the cooling system.

Table 6.4 provides the conductive losses of the main components and the total losses when operating at the converter's maximum output power with different input voltages. As you can see, the dominant losses occur in the output rectifier, T5–T8, and the power transformer, especially when the input voltage (Vin) is at its lowest.

For example, when Vin is 550 V, the combined power losses of the transformer and output transistors amount to 397 W. **It is important to note that the transformer and the rectifier's components need to be placed close to each other,** which means they must be as compact as possible. Achieving this while ensuring efficient cooling and minimizing the parasitic resistance (*R*—Not shown in Fig. 6.7)

Table 6.5 The conductive stress on the main passive power components

Vin	VC2res, RMS	IC2res, RMS	IC2, RMS	IC1, RMS	Tr VA	Pout/Tr VA	LV Tr RMS	LV Tr peak
550 V	445 V	26.3 A	13.8 A	433 A	22,828	0.52	587 A	1105 A
650 V	349 V	22.7 A	12.1 A	357 A	20,157	0.6	536 A	917 A
800 V	237 V	19.6 A	11.7 A	255 A	19,217	0.624	476 A	731 A

Table 6.6 Efficiency of the LLC Converter for different power levels and different input voltages

	Pout 6000 W		Pout 1200 W	
Vin	*f*com	Eff	*f*com	Eff
550 V	83 kHz	97.6%	87 kHz	92.9%
700 V	96 kHz	97.8%	98 kHz	94.1%
800 V	122 kHz	97.9%	125 kHz	94.8%

Table 6.7 The results and the overload capabilities of the LLC converter

Vin	Load	Pout	Eff	$I_{T2,RMS}$	I-LV Tr	f_{COM}	$I_{C2res,RMS}$	$I_{C1,RMS}$
550 V	0.06 Ω	14,152 W	95.5%	42.3 A	723 A	80 kHz	32.5 A	536 A
650 V	0.055 Ω	15,984 W	96.0%	42.9 A	735 A	91.7 kHz	30.4 A	504 A
800 V	0.04 Ω	22,126 W	95.6%	42.9 A	880 A	118 kHz	33.7 A	482 A

and inductance (L_1 and L_2 in Fig. 6.7) in the circuit is a significant challenge, both technically and financially. The duty cycle of power conversion (*P–C*) represents the percentage of cycle time that the topology uses to transfer energy through the transformer to the load. When the input voltage is 550 V, *P–C* is 56.5%, which is relatively low, and it can negatively impact the performance of the converter and its overall cost.

Table 6.5 shows the conductive stress on the main passive power components. We observe that the **Coefficient of Power Transformer Use** averages around 58%. The apparent power (VA) of the transformer is calculated by multiplying the RMS current on the primary side (i.e. the high-voltage side) of the transformer with the RMS voltage across the primary winding, including the leakage inductance of the transformer. One of the main issues highlighted in Table 6.5 is the high RMS current in the low-voltage side components when Vin is low, particularly when it is 550 V. This explains the drop in efficiency when the input voltage is low.

Table 6.6 provides the main performance characteristics of the LLC converter at different power levels, up to 50% of maximum power, with varying input voltages. Under these conditions, the performance is satisfactory, with no significant issues noted.

Table 6.7 shows the results and the overload capabilities of the LLC converter. These overload conditions were selected to push the converter components to their maximum tolerable stress levels. It is critical that the control and protection system be fast and sophisticated to handle these situations, ideally without interrupting the peak resonant current.

Figure 6.8 displays the simulation waveform of the output voltage and the timing when the load changes by a factor of 10, from 12 kW down to 1.2 kW, with Vin at 550 V—the simulation does not include an output filter. The output capacitor, C_1, is 1000 μF in our simulation, and this represents the worst-case scenario. The results show that this transition is manageable, with no damage to the output transistors or the ceramic output capacitor. For the load, the situation depends on its nature. If the primary load is a battery, which is often the case in this application, there should be no issues. For more sensitive loads, a filter is typically employed after this transient spike to significantly reduce overshoot. Alternatively, Metal-Oxide Varistors (MOVs) can be used, as they are widely popular in automotive applications.

Table 6.8 shows the delta changes in the control system needed to adjust the output voltage by 0.3 V, or 1% when the output power is at 12 kW. Upon inspection, we see that the change in frequency required to produce a 0.3 V change in the output voltage increases with the input voltage. This indicates that the LLC converter is relatively easier to control as its input voltage moves towards its highest voltage.

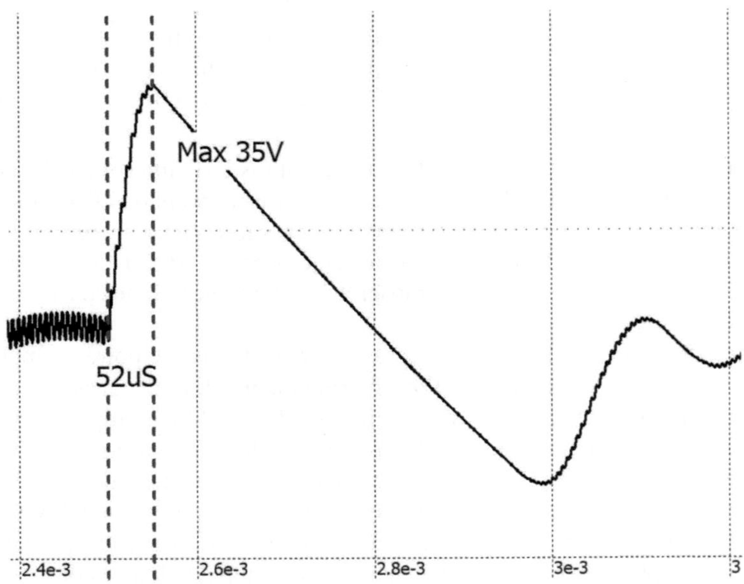

Fig. 6.8 The simulation of the output voltage and the timing when the load changes by a factor of 10

Table 6.8 The changes in commutation frequency that produce a change in Vout by 0.3 V when the output power is 12 kW

Vin	550 V	650 V	800 V	Average
Hz	Δ 650 Hz	Δ 819 Hz	Δ 1400 Hz	Δ 956 Hz
Percent	0.8%	0.9%	1.17%	0.957%

6.2.3.1 The Resonant LLC Converter's Complete Schematic

Now, let's consider the complete schematic of the LLC converter, including the basic topology and along with the schematics of the filter for both the high-voltage and low-voltage sides. Figure 6.9 shows the detailed schematic of the converter and its filters.

6.2.3.2 Low-Voltage Side

On the low-voltage side, handling high currents over 200 A (and peak currents reaching 700–800 A, as in our case) presents significant challenges in implementing a low-impedance circuit for the output capacitor, C1. Therefore, C1 is primarily selected based on its current and voltage rating, followed by its capacitance value. Given the high current demands of this application, an array of ceramic capacitors with Surface-Mounted Assembly (SMA) are the best choice. This decision significantly impacts the commutation frequency due to the additional inductance of the PCB and conductive losses.

The typical ripple voltage on the output connection of the power stage is a maximum of ±5% under full power. Differential inductors L3–L6 can be implemented

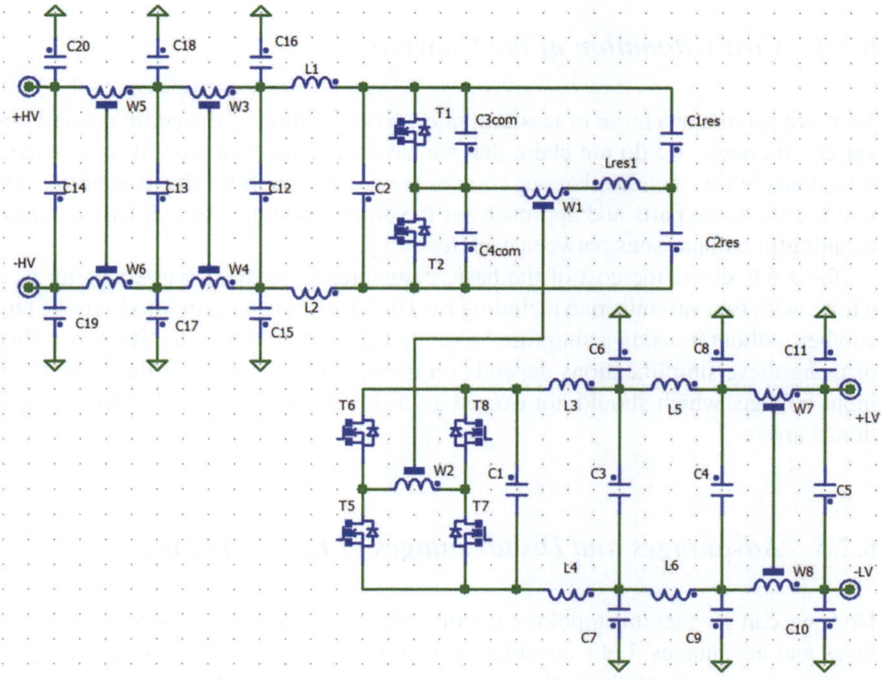

Fig. 6.9 The schematic of a Resonant LLC Converter with input and output filters

relatively easily using clamp cores, such as part number 00K4119U090, mounted on a busbar. The use of a busbar is essential in this setup to manage the 400 A current, which requires heavy conductive components. Capacitors C3 and C4 can be of various types, depending on their specific application requirements. A common-mode filter is necessary in this case because when the input voltage (Vin) drops, especially to 550 V, we experience high-frequency voltage ringing across transistors T5–T8, which occurs when T5–T8 are not conducting [10]. While clamping circuitry can be used to mitigate this problem, it complicates the low-voltage side of the power stage. If the input voltage range does not exceed a ratio of 1.3, a high-current common-mode choke might be unnecessary.

6.2.3.3 High-Voltage Side

The situation on the high-voltage side is quite similar. The high amplitude and relatively high *dV/dt* rates, up to 6–8 kV/μs, require the implementation of a common-mode filter. However, high-performance common-mode filters for the high-voltage side are often prohibitively expensive. A more cost-effective alternative would be to use a two-stage common-mode choke. This setup can achieve comparable or better performance to a single high-quality common-mode filter but at a reduced cost.

6.2.4 Cost Estimation of the Converter

Now, we can make a more or less accurate cost estimation of the entire power converter. Of course, we do not claim that we have made the best possible component selections, but we believe that our choices are good enough for this evaluation. We will use the same parts and approach for the other topologies so that we can make meaningful comparisons between each topology.

Table 6.9 shows the cost of the basic components for the power converter as a whole, with two variants: one including the High Current common-mode choke and another without it, using a simplified filter on the high-voltage side. The possibility of using these simplifications depends on the operational range of the converter's input voltage, which should not exceed 1.25–1.3 times its nominal value as mentioned above.

6.2.5 Advantages and Disadvantages of LLC Topology

Now, we can see that this topology is workable, but, of course, it has both disadvantages and advantages. Let's consider these points, starting with the disadvantages, because if we overlook them, they may eventually cause problems.

Table 6.9 The cost of basic components for power stage of a whole unit with high current common mode choke and without high current common mode choke and simplify filter of high voltage side

Parts	Part number	QTY	Price 1p $	Total $	Total Comm	Total $	Total without
T1, 2	UF4SC120030K4S	4	19.96	79.84		79.84	
T5, 6, 7, 8	IAUC120N06S5N032	40	0.73	29.2		29.2	
W1–W2	Power Tr without holders	1	180.0	180.0		180.0	
Ladd	E58/11/38	4	3.3	13.2		13.2	
C1res, C2res	R76UN1820JE00J	26	0.4	10.4		10.4	
C2	C4AQUBU4180A1XJ	4	0.99	3.96		3.96	
C1	CL32B106KBJNNNF	200	0.127	25.4		25.4	
Power supply for gate drivers T1–T2	AM2D-1218SZ	2	4.6	9.2		9.2	
Gate driver T1–T2	UCC53x0	2	0.91	1.82		1.82	
Power supply for gate drivers T5–T8	LMT78_15-1.0R	1	2.9	2.9		2.9	
Gate driver T5–T8	UCC5390-Q1	4	1.52	6.08		6.08	
Power stage	**Total**	**1**			**375.2**		**375.2**
C3com, C4com	R76UI056050H4J	10	0.2	2		2	
Aux circuits		**1**			**2**		**2**
L1, 2	CPEX3231A-270MC	2	11.8	23.6		23.6	
W3, 4, 5, 6	B82726S6243A040	2	9.4	18.8		N/A	
W3, 4	B82726S6243A040	1	9.4	N/A		9.4	
C12, 13, 14	C4AQUBU4180A1XJ	5	0.99	3.95		3.95	
L3, 4, 5, 6	00K4119U090	8	1.5	12.0		12.0	
C3, 4, 5	CL32B106KBJNNNF EMHS500ARA361MKE0S	30 6	0.127 0.65	3.81 3.9		3.81 3.9	
W7–W8	High current common mode choke	1	60	60		N/A	
Filters	**Total**	**1**			**126.06**		**56.66**
Power stage	Aux circuits, filters				503.26		433.86
Coefficient	On big size of parts	1.3			150.98		N/A
Coefficient	On big size of parts	1.2			N/A		86.77
	Total cost of converter				**654.24**		**520.63**

6.2.5.1 Disadvantages

The main disadvantage of this topology is its variable frequency control under full power. All other disadvantages are a consequence of this main issue. The variable frequency control under full power is an inherent characteristic of LLC topology, and we cannot change it. If we were to change this, we would essentially create a different topology, such as the Resonance with PWM Converter or another type. Some have tried to use this topology with a fixed commutation frequency [11–13], but this approach is very controversial, with the main criticism being that it is too complicated and increases costs. Let's consider the disadvantages that stem from this primary issue:

- **Increased size of power components**: The power transformer and output capacitors tend to be larger to handle the requirements of the system.
- **High stress on resonant capacitors**: Resonant capacitors experience a significant level of stress during operation.
- **Complex output filtering**: The need for a common-mode choke complicates the output filter design.
- **Reduced efficiency with lower input voltage**: Efficiency tends to decrease as the input voltage drops.
- **Higher costs**: Addressing these disadvantages often requires additional expenses, including more sophisticated components and control strategies.
- **Relatively slow response time**: The topology's ability to respond to changes in load or input conditions is limited.
- **Sophisticated control and protection needs**: Overload conditions require advanced and precise control mechanisms to ensure system stability and prevent damage.
- **Challenges in synchronous rectification**: Implementing effective synchronous rectification is difficult [14, 15].
- **Complex Soft Start**: Achieving a smooth startup sequence, particularly under load, is a challenge for this topology [16–20].

6.2.5.2 Advantages

Despite these challenges, the LLC topology also offers several notable advantages:

- **Soft switching across all load and voltage ranges**: The primary benefit of the Resonant LLC Converter is its ability to maintain soft switching across the entire range of load conditions and input voltages, which significantly improves efficiency and reduces stress on the components.
- **Cost-effective in systems with preregulators**: When used in applications with a **preregulator**, such as a Power Factor Correction (PFC) stage, the LLC topology becomes simpler and more cost-effective.

These advantages make the LLC topology a compelling choice for applications that demand high efficiency, especially in systems where the input voltage is relatively stable.

6.2.5.3 Operation and Application Area of LLC Topology

Now that we've reviewed the advantages and disadvantages of the LLC topology, let's consider its operation area and typical applications. The most common application area for LLC converters is in systems with high current on the low-voltage side, combined with a relatively low range of input voltage. This topology is also preferred in scenarios where it's relatively easy to achieve low electromagnetic interference (EMI). Table 6.10 provides an overview of these preferred operation and application areas. It's important to note that these values are not absolute; they have some tolerance depending on the specific target or requirements.

6.2.5.4 Modification

Many of the disadvantages have led to efforts to solve these problems. References [11–13, 21–23] show different approaches to improve this topology. We would like to contribute to these efforts by offering another solution. This topology has the potential to be modified by splitting the functions of the power transformer. In this approach, the power transformer primarily handles power conversion, while the charge of the resonance capacitor, when the input voltage (Vin) decreases, is managed by an additional inductor (Ladd). Figure 6.10 shows a schematic with the additional inductor, where Ladd = 40 μH, Lmag = 120 μH, and Lres1 = 3.2 μH. Tables 6.11 and 6.12 compare the simulation results of the conventional LLC converter (Conv LLC) with the modified LLC converter (Mod LLC) with different input voltages of 550 V and 800 V, respectively. We can observe that the modified converter

Table 6.10 Operation and application areas of the basic LLC topology

Power conversion (P-C)	56–89%	
Co/XMFR	52–62%	
Stab/coeff	0.8, 0.9, 1.17	0.957 average
Max commutation frequency	1000 kHz	Limited by passive power components
Start-up method	Special control	
Reliability	Middle	High stress of resonance capacitors
Input voltage max	850 VDC	Limited by semiconductors 1200 V
Input voltage min	30–200 VDC	Below 200 VDC increase cost
Input current	Max 100 A	Above 50 A increase cost of resonance capacitors
Range of vin	1.5 times	Above 1.25 start to suffer cost and efficiency
Load range	0–100%	Soft-switch full range
Output current	Around up 500 A	Limited by cost of implementation secondary side of power stage, include transformer
Response time	Middle	Limit by control, change frequency
Direction	Step-up or step-down	Unidirectional

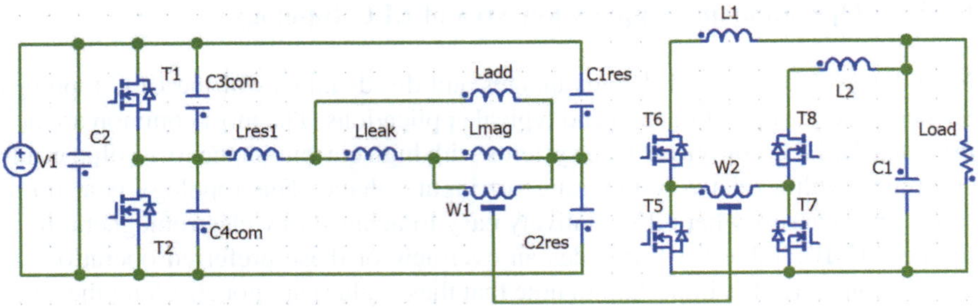

Fig. 6.10 The schematic of the Resonant LLC Converter with additional inductor (L_{add}) in parallel with the transformer

has higher efficiency, with a significant reduction in power losses in the output circuit and power transformer. This reduction in losses helps alleviate challenges in implementing this part of the converter. By minimizing losses, the output circuit and power transformer can be more compact, which also simplifies cooling. The placement of the additional inductor is not critical for this configuration.

Another modification involves changing the LLC topology to a CLLC topology. The CLLC configuration operates on the same principles as the LLC topology, sharing the same disadvantages and advantages. Compared to the LLC, the CLLC is inherently bidirectional. In other words, CLLC is a significant modification of the LLC topology but not a completely new topology. Therefore, we have not included CLLC in the list of topologies under consideration.

6.2.5.5 Conclusion

The LLC converter‘s apparent simplicity can tempt design engineers to select it without fully considering its implications. This tendency is evident from the numerous optimization efforts surrounding this topology. We urge design engineers to carefully evaluate the specific requirements of their application before choosing the LLC topology.

It’s also worth noting that the LLC topology was developed many years ago, during a period that preceded the significant advancements in power electronics. Nowadays, costs are largely driven by magnetic components rather than semiconductors, and this trend is likely to continue. The LLC topology has a specific area of application, where it is most suitable, and it is also ideally suited for WBG semiconductors.

6.3 Phase-Shift Topology Converter

The phase-shift topology is a type of isolated DC–DC power converter topology. It is primarily a soft-switching configuration that uses ZVS. The regulation of the output voltage in this topology is achieved by adjusting the phase angle

Table 6.11 Comparison of the simulation results for the conventional LLC and modified LLC when Vin = 550 V

	Pout W	T1–T2	T5–T8	Tr	T5–T8 +Tr	L add	Tr VA	Pout/Tr VA	Eff	P-C	f_{COM}
Conv LLC	11,840 W	55 W	207 W	190 W	397 W	n/a	22,828	0.52	96.3%	56.5%	81 kHz
Mod LLC	11,944 W	55 W	185 W	124 W	309 W	17 W	16,469	0.725	96.9%	63.9%	85 kHz

Table 6.12 Comparison of the simulation results for the conventional LLC and modified LLC when Vin = 800 V

	Pout W	T1–T2	T5–T8	Tr	T5–T8 +Tr	L add	Tr VA	Pout/Tr VA	Eff	P-C	f_{COM}
Conv LLC	12.0 kW	31 W	136 W	166 W	302 W	n/a	19,217	0.624	97.3%	89.3%	119 kHz
Mod LLC	11.9 kW	32 W	122 W	129 W	251 W	13 W	14,283	0.836	97.6%	98.7%	123 kHz

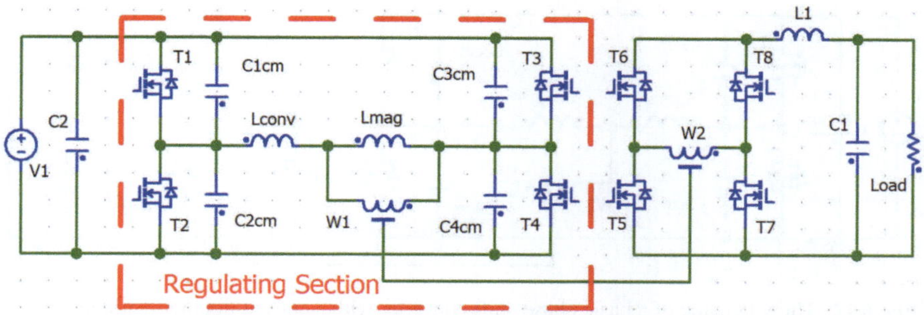

Fig. 6.11 The schematic of a basic phase-shift converter

commutation between the legs of the bridge power converter, which gives this topology its name. This approach has been in use since the late 1980s, and integrated circuits (ICs) for its control were developed soon after [24]. Today, there are numerous application notes available for developing and implementing this topology [25, 26].

Figure 6.11 shows the basic schematic of this topology. The output rectifier in this topology typically employs an inductive filter (L_1). There are several variations of this type of rectifier, and the choice of modification depends on the output current and voltage requirements. In our application, we will use a Current-Doubler Rectifier [27, 28], as illustrated in Fig. 6.12. This rectifier is particularly suitable for low-voltage (30 V) and high-current (400 A) applications. One of its key advantages is that it reduces the current through the secondary winding of the power transformer by approximately half compared to a bridge rectifier. This leads to a significant reduction in transformer size and improved efficiency, with an increase of about 0.25–0.75%.

However, the Current-Doubler Rectifier requires a higher number of turns in the output winding, as the output winding voltage is doubled. In our case, we will use two turns instead of one. Although this rectifier has some disadvantages, such as increased voltage ringing and the need for transistors rated for at least twice the voltage of those used in bridge rectifiers, the benefits outweigh these drawbacks. The alternative, a Center-Tapped Rectifier, is not considered in our design due to several reasons [29], the most significant being the complexity and expense of designing a center-tapped winding capable of handling 400 A.

6.3.1 Operation of Phase Shift Topology

To evaluate the operation of the Phase Shift topology, we will use the schematic from Fig. 6.12 under the following conditions:

- **Input voltage (V1)**: 700 V
- **Output voltage (Vload)**: 30 V

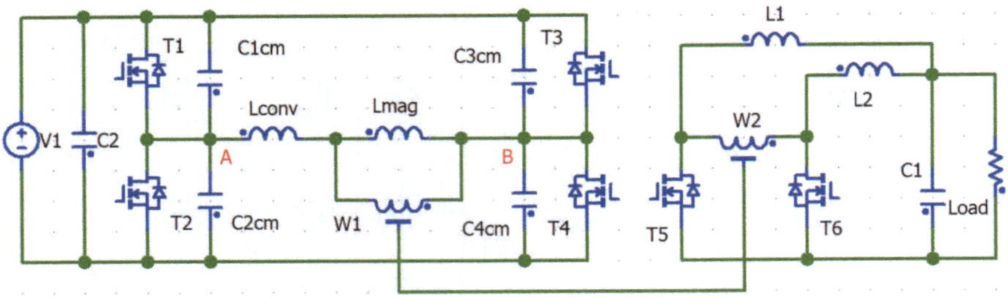

Fig. 6.12 The schematic of a basic phase-shift converter with a current-doubler rectifier

- **Maximum power**: 12 kW
- **Commutation frequency**: 100 kHz
- **Converter inductor (Lconv)**: 5 μH
- **Commutation capacitors (Ccom1–Ccom4)**: 0.75 nF each
- **Inductors (L1 and L2)**: 4.2 μH each
- **Output capacitor (C1)**: 200 μF
- **Power switches (T1–T4)**: UF4SC120030K4S
- **T5 and T6**: 10 EPC2034 devices connected in parallel
- **Power transformer's turns-ratio**: 15:2

All simulations were conducted under non-optimal conditions to demonstrate the basic implementation and potential of this topology. The output circuit was modeled as ideal, without simulating reverse recovery diodes and the ringing in output circuits. We will address this issue in the following sections. Figure 6.13 shows the simulation waveforms of the converter, while Figs. 6.14a–e depict the equivalent circuits for each operational mode.

Starting with Fig. 6.14a, which illustrates the time segment from t0 to t1. The device T1 turns off under ZVS at t0. The commutation time is determined by the current through L_{conv} and the values of the commutation capacitors, C_{com1} and C_{com2}. Once the commutation time concludes, the device T2 starts conducting in diode mode, with a small portion of the current returning to the source V1 and C2. This is a crucial point for this topology. In other words, the commutation time is a function of the commutation current, and if the commutation current is too low, hard switching will occur instead of soft switching. This implies that the *dV/dt* of the transistors is variable, and the dead time between the gate signals for T1 and T2, as well as T3 and T4, may be insufficient to ensure soft-switching commutation under low current conditions.

Moving to Fig. 6.14b, covering the time segment t1–t2, the device T2 turns on under ZVS conditions at t1, initiating a new cycle of power conversion during the period t1–t2. During the interval from t0 to t2, inductors L1 and L2 discharge to the load through devices T5 and T6. At t2, as shown in Fig. 6.14c, the current through device T6 reaches zero, and the complete process of power conversion begins, with energy transferring from the source to the load and accumulating in inductor L2. Meanwhile, inductor L1 continues discharging to the load. At t2, another critical point occurs: device T6 turns off under hard-switching conditions. For this illustration, ideal diodes are used in devices T5 and T6, without considering recovery time

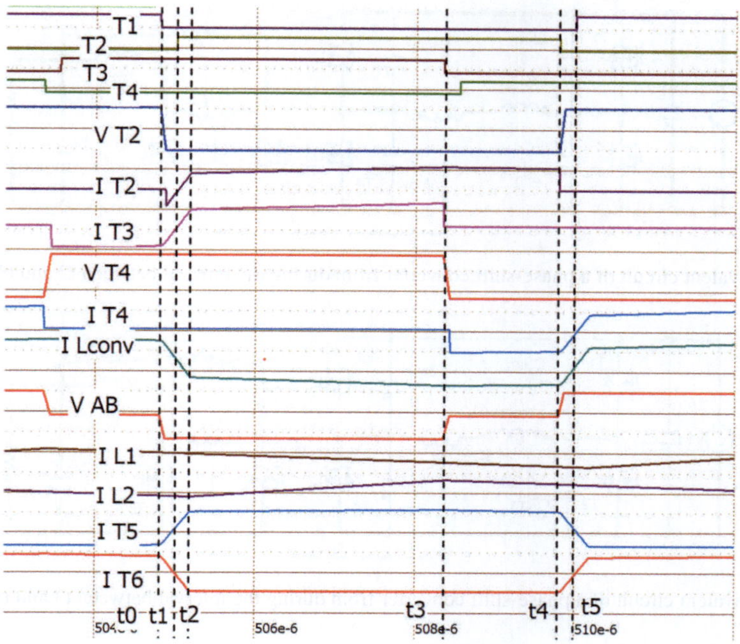

Fig. 6.13 The simulation waveforms of the phase-shift converter in Fig. 6.14

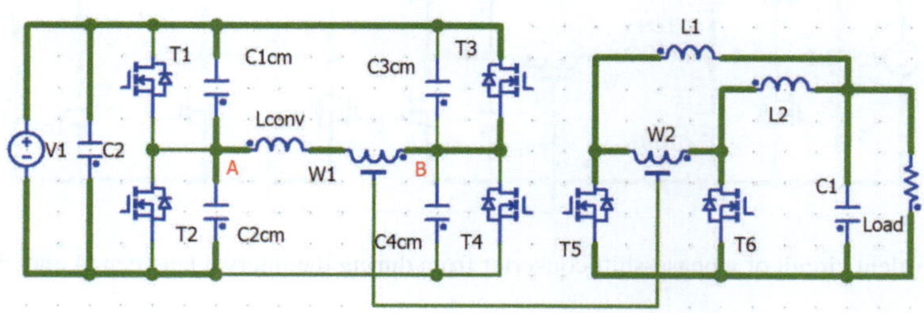

Fig. 6.14 (**a**) Equivalent circuit of a phase-shift converter from during the interval between t0 and t1.

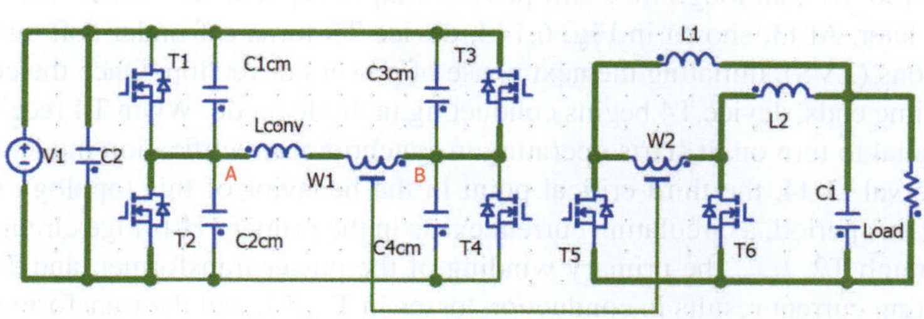

(**b**) Equivalent circuit of a phase-shift converter from during the interval between t1 and t2

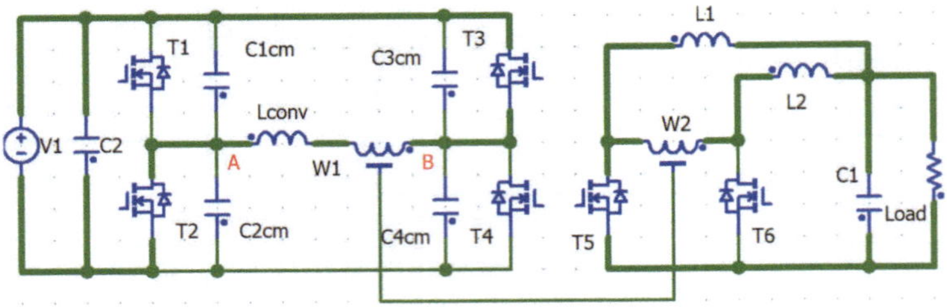

(**c**) Equivalent circuit of a phase-shift converter from during the interval between t2 and t3

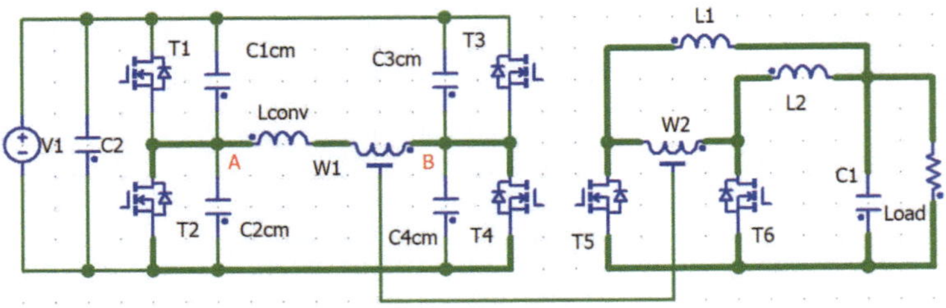

(**d**) Equivalent circuit of a phase-shift converter from during the interval between t3 and t4

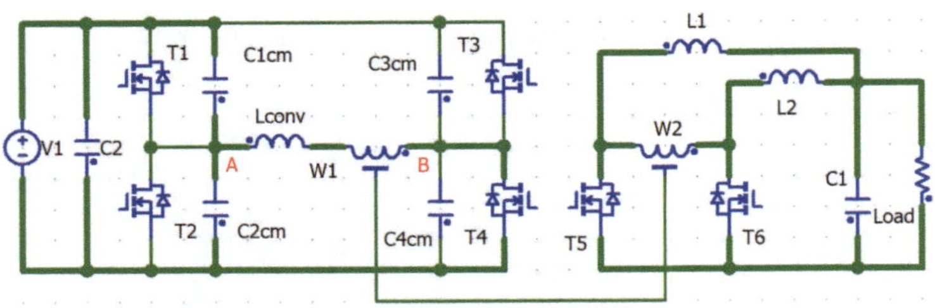

(**e**) Equivalent circuit of a phase-shift converter from during the interval between t4 and t5

or output capacitance. In reality, when body diodes have a recovery time, energy recovery issues can arise. Therefore, GaN transistors were chosen for devices T5 and T6 [30, 31], although they still possess output capacitance, which will be discussed later. At t3, shown in Fig. 6.14d, device T3 turns off under soft-switching conditions (ZVS), initiating the next phase of power conversion. Once the commutation time ends, device T4 begins conducting in diode mode. When T4 receives the gate signal to turn on, it starts operating in synchronous rectification mode. During the interval t3–t4, the third critical point in the behavior of this topology occurs. During this period, a circulating current exists in the primary H-bridge circuit, flowing through T2, L_{conv}, the primary winding of the power transformer, and T4. This circulating current results in conduction losses in T2, T4, and the transformer without contributing to power conversion. Meanwhile, L1 and L2 continue to discharge to the load through devices T5 and T6. At t4, shown in Fig. 6.14e, device T2 turns off under ZVS conditions, initiating a new half-cycle that mirrors the half-cycle between t0 and t4.

6.4 Simulation Results Under Varying Input Voltages and Load Levels

We now consider the results of simulations under different conditions, varying both the input voltage and load levels. Figure 6.15 shows the schematic used for this purpose, closely resembling the actual implementation of a phase-shift topology. As seen in the schematic, additional components have been incorporated to enhance performance and address potential issues.

1. **Snubber Circuit (Csnub and Rsnub)**
   We introduced a snubber circuit consisting of Csnub (30 nF) and Rsnub (1.5 ohms) to mitigate and reduce ringing, which occurs at time t2 (as shown in Fig. 6.14c). There are various methods to address this issue, but we opted for the simplest and most cost-effective solution. While more sophisticated approaches may yield better results, they often increase both cost and circuit complexity [32, 33].
2. **Soft-Switching Commutation Circuit (L3, L4, C3–C6, D1–D4)**
   The inductors L3 and L4 (170 μH each) and capacitors C3–C6 (5 μF each) were added to enable soft switching under low-load conditions by providing additional current for commutation. Diodes D1–D4 prevent ringing in this circuit.
3. **DC-Bias Mitigation (Cbl)**
   The capacitor Cbl (20 μF) was included to prevent the power transformer from saturating. Due to the inherent design of H-bridge topologies, DC bias is always present—though the magnitude can vary. In our case, the presence of a circulating current increases the DC bias.
   Finally, this topology achieves full soft switching under the simulated conditions. Therefore, for loss calculations, we considered only the conductive losses of the main power components. It is important to note that without these additional components, the topology would not deliver acceptable performance.

6.4.1 *Analysis of Conductive Losses and Performance Metrics*

Table 6.13 presents the conductive losses of the main components and the total losses at maximum output power. As can be seen, the primary sources of losses are the output rectifier and the power transformer.

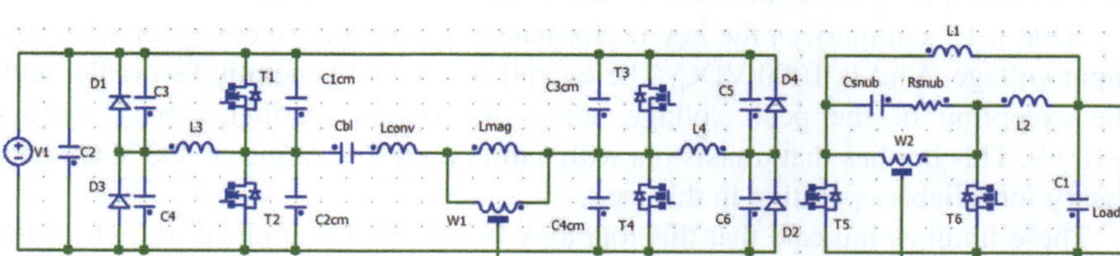

Fig. 6.15 Schematic of the phase-shifted topology used in the simulations and calculations

Table 6.13 Conductive losses of main components and total losses of phase-shift converter

Vin	Pout	T1–T4	T5–T6	Tr	L1–L2	Rsn	L3 L4	Total W	Eff	P-C
550 V	11,920 W	57 W	122 W	89 W	43 W	30 W	6 W	347 W	97.2%	85.4%
700 V	12,000 W	63 W	120 W	93 W	43 W	47 W	7 W	373 W	97.0%	67.4%
800 V	12,000 W	63 W	117 W	94 W	43 W	61 W	8 W	386 W	96.9%	59.6%

Table 6.14 Switching stress and conductive stress of the main power components

Vin	Ioff-T2	Sw T2	Ioff T4	Sw T4	Vp T5–T6	Tr-VA	Pout/Tr VA	I-LV Tr
550 V	34 A	68 ns	39 A	67 ns	107 V	14,045	0.849	193 A
700 V	34 A	84 ns	40 A	78 ns	135 V	16,238	0.734	190 A
800 V	34 A	95 ns	41 A	85 ns	153 V	17,417	0.689	187 A

Table 6.15 Main performance of converter higher levels of Vin

Vin	Pout	Tr-VA	Pout/TrVA	Vp T5–T6	Eff	P-C
800 V	12,000 W	17,417	0.689	153 V	96.9%	59.6%
1000 V	12,000 W	19,284	0.622	189 V	96.7%	48.6%

Table 6.16 Phase-shift converter's performance at different power levels and varying input voltages

	Pout 6000 W					Pout 1200 W				
Vin	Ioff-T2	Sw T2	Ioff T4	Sw T4	Eff	Ioff-T2	Sw T2	Ioff T4	Sw T4	Eff
550 V	21.7 A	110 ns	25.7 A	99 ns	97.8%	11.9 A	230 ns	14.9 A	172 ns	95.7%
700 V	21.8 A	140 ns	27.4 A	110 ns	97.5%	12.8 A	250 ns	16.3 A	182 ns	95.2%
800 V	22 A	163 ns	28.2 A	122 ns	97.3%	13.4 A	260 ns	17.2 A	193 ns	94.5%

Table 6.14 details the switching and conductive stress on the main power components. It is evident that as the input voltage increases, the peak voltage across the output rectifier also rises. Notably, the power transformer achieves an efficiency above 99%. This high efficiency is partly due to the use of two turns on the low-voltage side, which effectively halves the current through the low-voltage winding compared to a single-turn configuration. Additionally, the current waveform contributes to reducing the conductive losses in the power transformer.

Table 6.15 summarizes the key performance metrics of the converter when the input voltage (Vin) is 1000 VDC. The overall results are generally favorable, with the exception of the peak voltage across the output rectifier, which reaches −189 V. This implies that transistors with a minimum Vdc rating of 300 V are necessary for reliable operation in this case.

These findings indicate that this topology is suitable for applications where the input voltage varies by up to 1.6 times. Table 6.16 shows the results at different

Table 6.17 Result under overload conditions

Vin	Load ohm	Pout	Eff	Ioff T2	Sw T2	Ioff T4	Sw T4
550 V	0.0375 Ω	19,586 W	96.1%	56 A	36 ns	60.4 A	41 ns
800 V	0.0375 Ω	24,486 W	95.1%	60 A	47 ns	69.5 A	48 ns

power levels and varying input voltages. It is evident that under light load conditions, the commutation current significantly decreases, and the switching time increases. We performed simulations with additional commutation inductors (L3–L4), which ultimately achieved soft-switching. However, it is crucial for design engineers to be aware that under light load, the system may experience hard switching instead of the intended soft switching.

6.4.2 Performance Under Overload and Transient Conditions

Table 6.17 presents the results under overload conditions. In this case, the switching time is very short, resulting in a high *dV/dt* of 15–16 kV/μs. This rapid rate of voltage change poses serious concerns for the converter's reliability and electromagnetic interference (EMI) performance.

While increasing the value of the commutation capacitors could mitigate this issue, it introduces challenges under light load conditions, requiring the design engineer to strike a careful balance.

Another critical scenario occurs when the full load is rapidly disconnected, known as load dump mode. Figure 6.16 illustrates the simulation of this condition, where the load was abruptly reduced by a factor of 10—from the full load (12 kW/0.075 ohms) to 1.2 kW/0.75 ohms. The capacitor C1 (1000 μF) is used without any additional filtering, resulting in a significant overshoot of 50%. This equates to a voltage spike of 45 V, which can be problematic, particularly if the load is sensitive or if ceramic capacitors are used in the output filter. If the load includes a battery, this overshoot might be somewhat acceptable, but it still requires careful attention.

Table 6.18 shows the control adjustments required to achieve a 0.3 V change in Vout when the output power is 12 kW.

6.4.3 Cost Estimation of the Power Stage

We can now make a more accurate estimation of the cost of the entire power stage. Figure 6.17 illustrates the complete power converter, including the input and output filters. As mentioned earlier, this topology exhibits a high level of overshoot in the output rectifier, which propagates directly to the output circuit. One reason this

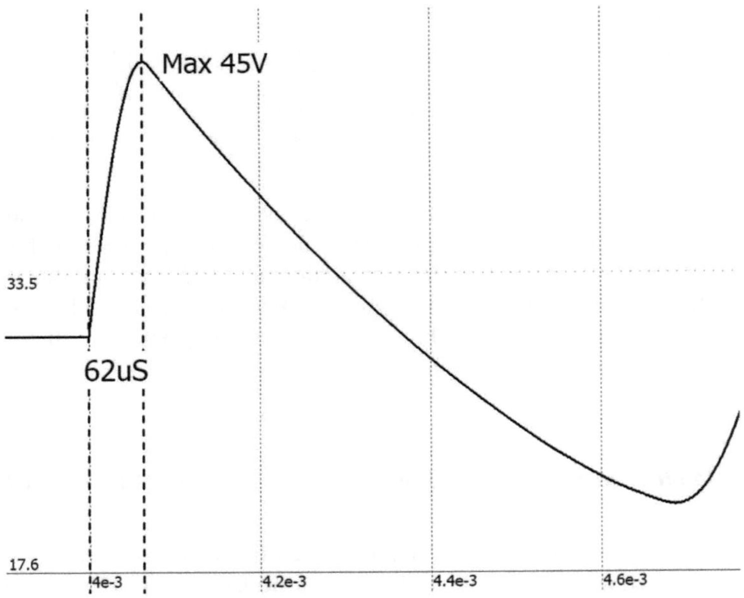

Fig. 6.16 Phase-shifted converter's output voltage and timing when the level of load changes by 10×—from full load (12 kW @ 30 V and 0.075 Ω) to 10% of the rated load (1.2 kW @ 30 V and 0.75 Ω)

Table 6.18 The changes in phase angle, switching time, and percentage of control to change Vout by 0.3 V when the output power is 12 kW

Vin	550 V	700 V	800 V	Average
Degree	1.8°	1.44°	1.15°	1.46°
Switching time	50 ns	40 ns	32 ns	41 ns
Control percentage	1.0%	0.8%	0.64%	0.81%

occurs is through the interwinding capacitance of the inductors in the output filter. To significantly reduce this issue, we recommend using two inductors connected in series instead of one. In this case, the capacitance between the input and output of the inductor will be reduced by at least half. Additionally, this configuration helps improve the cooling of the output inductor. Unfortunately, this design generates a high level of common mode noise, necessitating the use of a common mode filter. On the other hand, the need for a differential filter is not as critical when using a rectifier with an inductor, which is an advantage of this type of rectifier. Ultimately, the high-current common mode filter, output inductors, and RC snubber significantly impact the overall cost of the converter. While we don't claim that our component selections are the absolute best, they are reasonably effective, and we will use the same components for other topologies. Table 6.19 provides the cost breakdown of the basic components for the power stage of the entire unit.

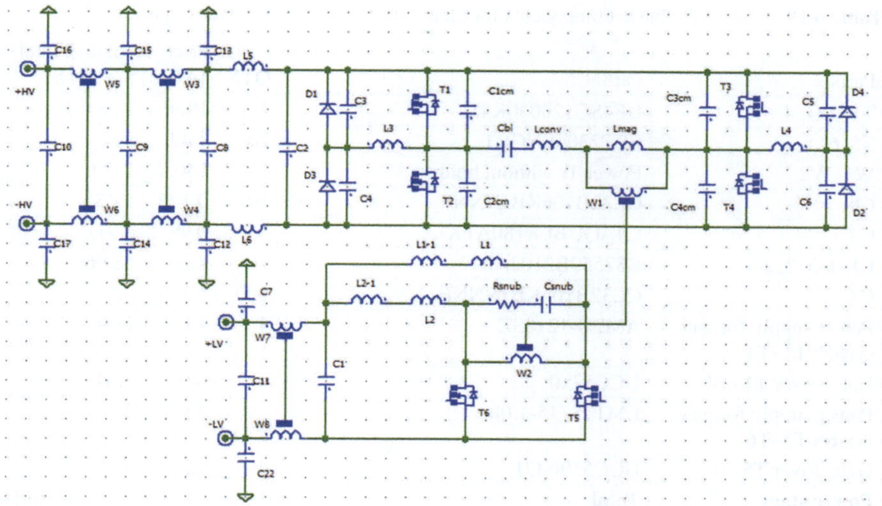

Fig. 6.17 The schematic of the complete phase-shift converter including input and output filters, snubbers, and components for soft-switching

6.4.4 *Advantages and Disadvantages of the Phase-Shift Converter*

Just like the LLC Converter, this topology is workable, but, of course, it has both disadvantages and advantages. Let's consider these points, starting with the disadvantages.

6.4.4.1 Disadvantages of the Phase-Shift Topology

Now, let's consider the disadvantages of this topology. **The primary drawback lies in the output rectifier,** which warrants further explanation. The rectifier operates with hard switching, where the output capacitance of the rectifier's diodes or transistors generates additional ringing during transients. Additional loss occurs because of the semiconductors' reverse recovery charge. These two conditions create ringing that increases electromagnetic interference (EMI) and adds stress to the output semiconductors due to the elevated voltage levels. GaN devices offer an advantage here since they reverse recovery charge is negligible and their output capacitance is relatively small.

Another disadvantage is the inclusion of an inductor in the output filter. This significantly impacts the overall cost and introduces additional conductive and core losses. While our calculations for inductor losses only consider the DC resistance (R_{dc}), in reality, the losses can be twice as high. The voltage stress across the rectifier semiconductors is directly dependent on the input voltage—higher input voltage results in higher voltage stress.

Table 6.19 Cost of the basic components for phase-shift converter's power stage

Parts	Part N	QTY	Price 1p $	Total $	Total $
T1, 2, 3, 4	UF4SC120030K4S	4	19.96	79.84	
T5, 6	EPC2304ENGRT	20	4.47	89.4	
W1–W2	Power Tr without holders	1	140.0	140.0	
Cbl	CL32B106KBJNNNF	10	0.127	1.27	
C2	C4AQUBU4180A1XJ	3	0.99	2.97	
L1, 1-2, 2, 2-1	B82559B2102A027	16	4.31	68.96	
C1	CL32B106KBJNNNF	100	0.127	12.7	
Power supply for gate drivers T1–T4	AM2D-1218SZ	4	4.6	18.4	
Gate driver T1–T4	UCC53x0	4	0.91	3.64	
Power supply for gate drivers T5–T6	LMT78_15-1.0R	1	2.9	2.9	
Gate driver T5–T6	UCC5390-Q1	4	1.52	6.08	
Power stage	**Total**	**1**			**426.16**
C1com, C2com C3com, C4com	R76UI122050H4J	4	0.2	0.8	
L3, 4	PFC3819QM-151K11B-00	2	11.4	22.8	
C3, 4, 5, 6	C4AQHLU4500P12J	4	1.57	6.28	
Rsnub	PWR163S-25-5R60F	4	1.75	7.0	
Csnub	CKC18C752MCGAC7210	4	0.8	3.2	
L5, 6	CPEX3231A-270MC	2	11.8	23.6	
Aux circuits	**Total**	**1**			**63.68**
C8, 9, 10	C4AQUBU4180A1XJ	4	0.99	3.96	
W3, 4, 5, 6	B82726S6243A040	2	9.4	18.8	
C11	CL32B106KBJNNNF EMHS500ARA361MKE0S	30 2	0.127 0.65	3.81 1.3	
W7–W8	High current common mode choke	1	60	60.0	
Filters	**Total**	**1**			**87.87**
Summary	Power stage	Aux	Filters		**577.71**
Coefficient	On big size of parts	1.5			**288.85**
	Total cost of converter				**866.56**

A further challenge is achieving soft switching across the full power range on the input bridge. In this topology, the switching time varies with the load: higher load results in shorter switching time, while lower load leads to longer switching time. This variability can exacerbate problems, particularly during transient conditions. These issues impose significant limitations on the application range of phase-shift topology.

The Phase-Shift topology is relatively old, having been in active use for roughly 40 years. Over this time, many solutions have been proposed to address or mitigate its inherent problems [34, 35], among others. However, these solutions still belong to the phase-shift family, meaning that fully resolving these issues likely requires developing an entirely new topology.

Another, less critical, disadvantage is the presence of circulating current during the t3–t4 interval (Fig. 6.14d) and the DC bias current through the primary winding of the power transformer. While the DC bias issue is relatively easy to address, managing the circulating current is more challenging [36].

6.4.4.2 Advantages of the Phase-Shift Topology

On the positive side, this topology offers simple control, with readily available ICs that provide good performance. It also achieves zero-voltage switching (ZVS) in the primary H-bridge and typically does not require an additional inductor, as the leakage inductance is often sufficient.

6.4.4.3 Operation and Application Area of Phase-Shift Topology

Now that we have clarified the advantages and disadvantages of the phase-shift topology, let's consider its application areas. The primary application for this topology is in circuits where rectifier diodes with minimal reverse recovery time and low output capacitance are required. This suggests that for optimal results, it is advisable to use this topology without synchronous rectification.

Table 6.20 outlines the operation and application areas for the phase-shift topology. It is important to note that the values provided are not absolute; they have tolerances depending on the specific targets. The authors have previously applied this

Table 6.20 Operation and application areas of the basic phase-shift topology

Power conversion	60–85%	
Co/XMFR	69–85%	
Stab/coeff	1, 0.8, 0.64	0.81 average
Max commutation frequency	300 kHz	Limited by implementation low voltage power stage hard switch rectifier
Start-up method	Itself with snubber	Without snubber, start hard switch
Reliability	Middle	Hard switch under overload
Input voltage max	750 VDC	Limited by semiconductors 1200 V
Input voltage min	30–100 VDC	Below 100 VDC increase cost
Input current	Max 100 A	Limited by commutation time
Range of Vin	1.8 times	Limited by cost of implementation secondary side of power stage, include transformer
Load range	30–100%	With soft-switch, by itself
Output current	Around up 600 A	Limited by cost of implementation secondary side of power stage, include transformer
Response time	Middle	Limit by output inductor
Direction	Step-up or step-down	Unidirectional

topology in an 18 kW isolated DC–DC converter with a 350 VDC output, incorporating auxiliary commutation inductors and rectifiers with snubbers, as described in [33]. However, this was done many years ago.

6.5 Dual Active Bridge Topology

The Dual Active Bridge (DAB) converter belongs to the family of isolated DC–DC bidirectional power converters. It is also sometimes referred to as a topology with an active rectifier. This topology was patented in 1991 [37], with its first publication in 1988 [38]. The DAB converter consists of two active bridges—one on the primary side and one on the secondary side of a high-frequency transformer. Figure 6.18 shows the basic schematic of this topology.

Generally, the regulation of a DAB converter is achieved through the mutual interaction of two bridges positioned on either side of the power transformer. This regulation can be implemented using various control methods, including phase-shift modulation, PWM, variable frequency, or combinations of these approaches. The DAB operates with semi-soft switching commutation, meaning it includes both a soft-switching region and a hard-switching region [39, 40]. The region in which the converter operates is influenced by the load level and input voltage.

This topology is the leading contender in the competition to be an isolated DC–DC converter the most modifications and optimizations. For example, details on optimization and calculations can be found in references [41–45]. For more information on modifications and optimizations, consult books [46, 47] or the following dissertation [48].

Before diving into the operation of the DAB converter, it's important to note that **a high number of modifications and optimizations typically indicate underlying issues with the topology**. This often suggests that design engineers are addressing the effects, or consequences, of a problem rather than the root cause of the problem. In the following sections, we will focus on the root cause.

For simulation purposes, we will use the basic phase-shift control method. Figure 6.19 presents the schematic used for the simulation.

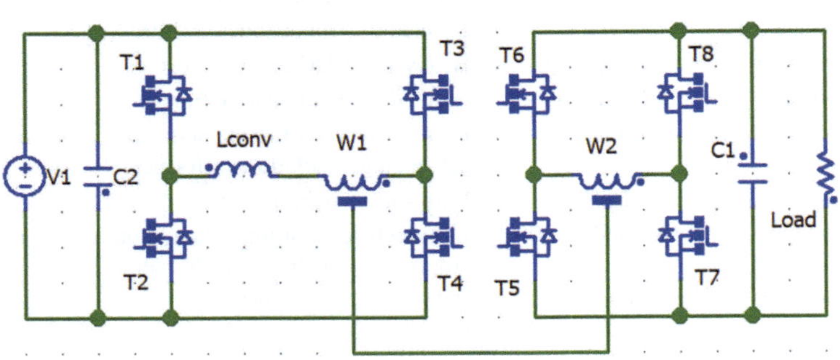

Fig. 6.18 The schematic of a basic dual-active bridge (DAB) converter

6.5.1 Operation of DAB Topology in Step-Down Mode

In this section, we analyze the DAB converter operating in step-down mode with the following parameters:

- **Input voltage (V1)**: 550 V
- **Output voltage (Vload)**: 30 V
- **Maximum power**: 12 kW
- **Commutation frequency**: 100 kHz
- **Inductor (Lconv)**: 29 μH
- **Commutation capacitors (C1com–C4com)**: 2 nF each
- **Commutation capacitors (C5com–C8com)**: 1 μF each
- **Output capacitor (C1)**: 1000 μF
- **Power switches (T1–T4)**: UF4SC120030K4S
- **Power switches (T5–T8)**: IAUC120N06S5N032ATMA1 (10 connected in parallel)
- **Turn ratio of power transformer**: 20

6.5.2 Design Considerations

1. **Turn Ratio Selection**
   The turn ratio of the power transformer was selected primarily to minimize the commutation current on the low-voltage side bridge. A turn ratio of 20 was chosen to achieve this goal, balancing the requirements of voltage step-down and component stress.
2. **Component Choices**
   - **Inductor (Lconv)**: 29 μH to manage power transfer at the given commutation frequency
   - **Commutation capacitors**: C1com–C4com at 2 nF each and C5com–C8com at 1 μF each, to provide soft switch commutation
   - **Output capacitors**: 1000 μF to maintain voltage stability at the output

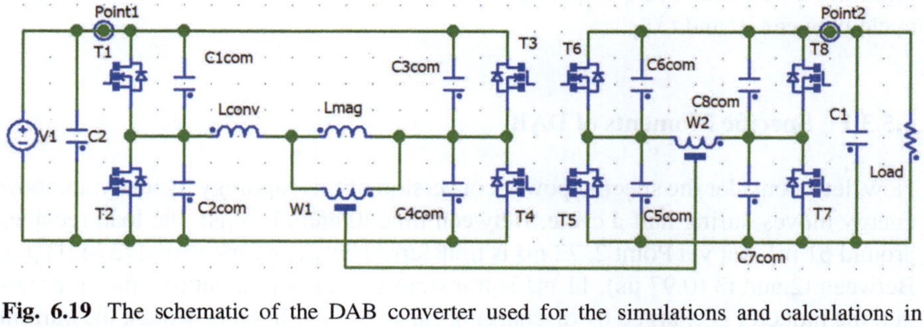

Fig. 6.19 The schematic of the DAB converter used for the simulations and calculations in this section

3. **Simulation Conditions**
   Simulations were conducted under non-optimal conditions to demonstrate the basic implementation and potential of the DAB topology. These simulations serve to illustrate the fundamental operation and feasibility of the design, despite not being optimized for peak performance. The output circuit is modeled as ideal, excluding the simulation of the PCB inductance. The underlying causes will be discussed further below.

6.5.3 Description of Operation

The analysis provides a clear understanding of the DAB converter's operation in step-down mode. The chosen turn ratio and component specifications reflect an initial design aimed at showcasing the converter's capabilities.

Figure 6.20 shows the simulation waveforms of the converter in step-down mode, while Fig. 6.21a–d illustrate the equivalent circuits for each operational mode.

Starting with Fig. 6.21a, which illustrates the time segment from t0 to t1: At t0, devices T7 and T6 turn off under ZVS, and devices T5 and T8 begin conducting in diode mode. At this moment, the current through Point2, as shown in Fig. 6.19, changes direction and starts to decrease as T5 and T8 conduct from t0 to t1. Additionally, energy from the power source, plus energy previously accumulated in Lconv, begins to transfer to the load and charge the output capacitor C1. At t1, devices T2 and T3 turn off under ZVS conditions. The current at Point1, as shown in Fig. 6.19, starts to decrease and changes direction during the switching period of devices T1–T4.

In Fig. 6.21b, at t1, part of the energy remaining in L_{CONV} begins to charge the input capacitor C2, while another part continues to move to the output circuit. By t2, the currents through Lconv, Point1, and Point2 reach zero. During the period from t2 to t3, the process of charging Lconv starts, as shown in Fig. 6.21c. This process involves the power source and capacitor C2 on the high-voltage side and capacitor C1 on the low-voltage side. In other words, capacitor C1 discharges in two directions: one to the load and the other to the inductor L_{CONV}. During this period, devices T5 and T8 remain in transistor mode. At t3, devices T5 and T8 turn off under ZVS, initiating a new half-cycle of power conversion that is a mirror image of the half-cycle between t0 and t3.

6.5.3.1 Specific Moments of DAB

Now, let's consider the specific power conversion of this topology by analyzing how energy moves during half a cycle. Between time t0 and t3 (5 μs), the load receives around 61 mJ, but via Point 2, 72 mJ is transferred between time t0 and t2 (4.03 μs). Between t2 and t3 (0.97 μs), 11 mJ is transferred via Point 2, but to a different target: the inductor is charged by discharging capacitor C1. In other words, the current through Point 2 changes direction.

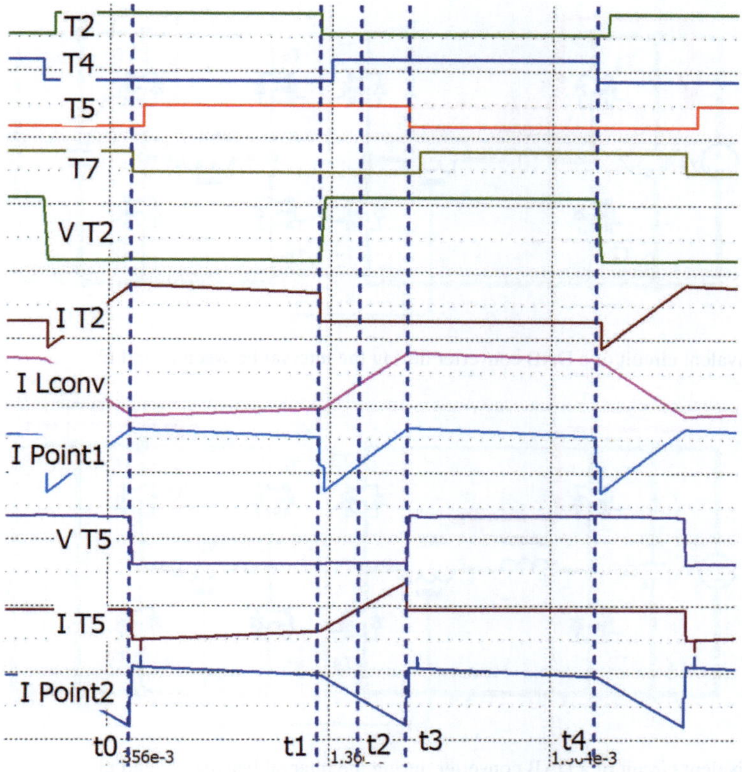

Fig. 6.20 The simulation waveforms of the dual active bridge converter in step-down mode

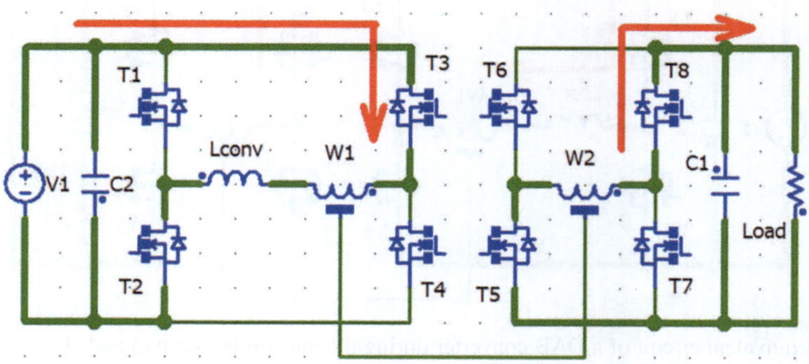

Fig. 6.21 (**a**) Equivalent circuit of a DAB converter during the interval between t0 and t1

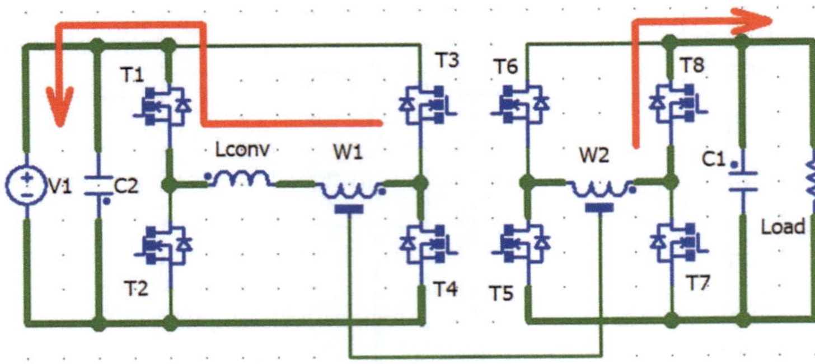

(**b**) Equivalent circuit of a DAB converter during the interval between t1 and t2

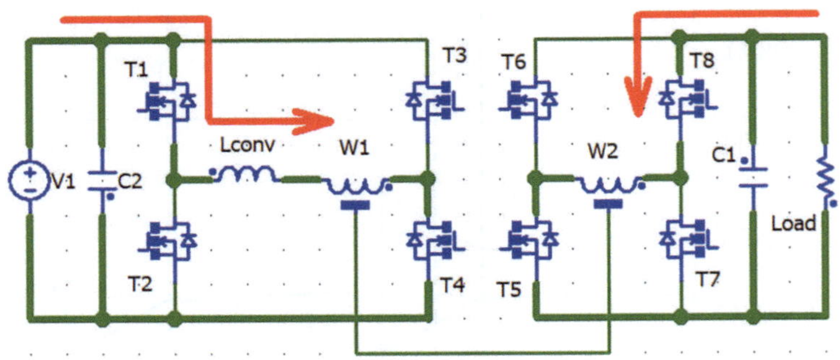

(**c**) Equivalent circuit of a DAB converter during the interval between t2 and t3

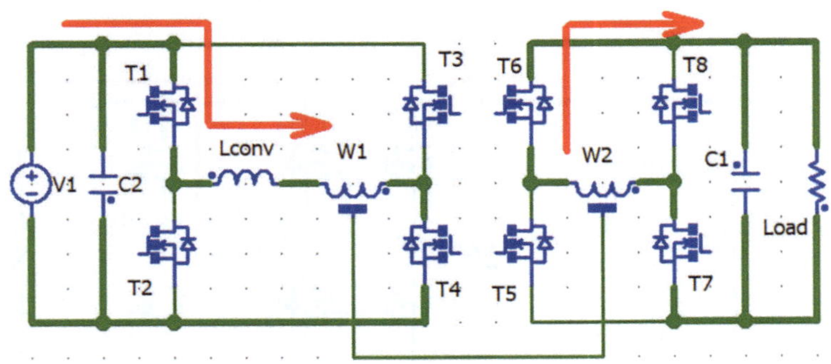

(**d**) The equivalent circuit of a DAB converter during the interval between t3 and t4

The source provides 70 mJ through Point 1 during the intervals t0–t1 and t2–t3, totaling 4.3 μs. However, between t1 and t2 (0.7 μs), 5.4 mJ is returned from the inductor to the source. These values correspond to a 12 kW converter operating at 550 V to 30 V.

This analysis suggests that **the DAB topology violates a fundamental principle of power conversion: energy should flow in one direction, from the source**

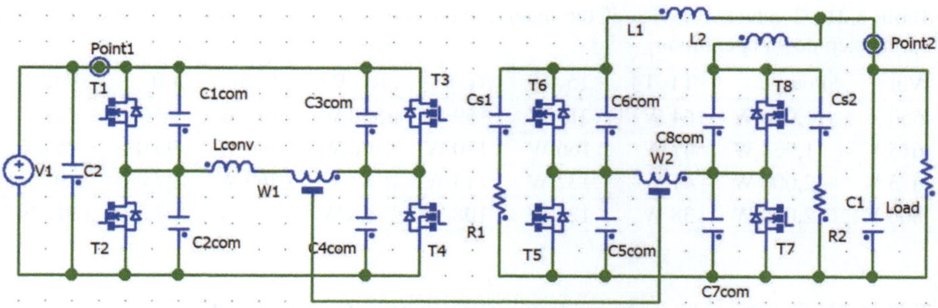

Fig. 6.22 A simplified schematic of the DAB converter with PCB inductances and an RC snubber

to the load. The consequence of this behavior is that at times t0 and t3, the current through Point 2 rapidly changes direction—from approximately 780 A in one direction to 780 A in the opposite direction. This results in a peak-to-peak current swing of 1560 A within a short switching period.

Another point to consider is the output circuit used in our simulation to demonstrate the principle of DAB operation. The circuit is idealized, without accounting for the inductance of the PCB. In real-world applications, such as the one we are considering, the PCB inductance is typically around 3–5 nH. Achieving this low inductance is challenging and costly.

Furthermore, the rapid change in current direction with such a high amplitude can lead to significant overshoot, which necessitates the use of a snubber circuit. However, discussing the design and implementation of such a circuit is beyond the scope of this book.

6.5.3.2 Simulation and Calculation

For the next set of simulations and calculations of the DAB converter, we used a simple RC snubber circuit. Figure 6.22 shows a simplified schematic of the converter that includes the PCB inductance, L1 and L2, which are 5 nH each. There is also an RC snubber where the capacitors Cs1 and Cs2 are 10 μF, and the resistors R1 and R2 are 0.2 Ω.

We then analyzed the simulation results under different input voltage conditions. This simulation was conducted under soft-switching conditions (ZVS), which do not accurately reflect real-life scenarios. Therefore, in our loss calculations, we only considered the conductive losses of the main power components. In real-world applications, additional losses from the PCB, busbars, and filters must be included, which could add significant losses, ranging from 2% to 4%.

In this context, duty cycle of power conversion (*P-C*) refers to the period when the energy at Point 2 is transferred to C1 and the load. Tables 6.21 and 6.22 present the results of this simulation and calculation.

Analyzing these results, it is evident that the basic DAB topology is not suitable for this application. The applicable input voltage range for this topology is

Table 6.21 Conductive losses of the main components and total losses of the DAB converter during step-down operation

Vin	Pout	T1–T4	T5–T8	Tr	R1–R2	Total	Eff	P-C
550 V	11,960 W	64 W	215 W	169 W	350 W	798 W	93.7%	78%
615 V	11,968 W	47 W	166 W	130 W	166 W	499 W	96.0%	85.4%
675 V	12,000 W	41 W	132 W	113 W	83 W	369 W	97.0%	89%
800 V	12,000 W	38 W	121 W	108 W	68 W	335 W	97.3%	91.2%

Table 6.22 Switching stress and conductive stress main power components

Vin	Ioff-T2	Ioff T5	I-C1 RMS	I-C2 RMS	Tr-VA	Pout/Tr VA	I-LV Tr RMS
550 V	29 A	771 A	465 A	16.9 A	17,110	0.7	580 A
615 V	27.4 A	568 A	325 A	14.1 A	14,700	0.86	500 A
675 V	28.5 A	409 A	244 A	13.4 A	13,524	0.89	460 A
800 V	34 A	45 A	196 A	15.3 A	13,350	0.9	448 A

Table 6.23 The delta changes of control for change Vout on 0.3 V when output power is 12 kW

Vin	550 V	615 V	675 V	800 V	Aver
Degree	1.44°	1.08°	1.08°	0.72°	1.08°
Switching time (ns)	40 ns	30 ns	30 ns	20 ns	30 ns
Percentage	0.8%	0.6%	0.6%	0.4%	0.6%

between 615 and 800 V. However, determining which modifications of the DAB topology would be appropriate for this application is beyond the scope of this book. Therefore, we will not consider other operating modes, such as low load and load dump.

Table 6.23 shows the changes in degrees, time intervals, and percentage of phase shift that are required to adjust Vout by 0.3 V when the output power is 12 kW and Vout is 30 V.

6.5.3.3 Step-Up Mode

The DAB topology is inherently bidirectional. Let's consider this topology in step-up mode under the following conditions: Vin ranges from 24 to 30 V, with an output around 675 V, while the input current is limited to 400–430 A. This current limit is a requirement of the battery, as the battery or special charger used as the source cannot exceed this output current.

Figure 6.23 shows the schematic of the topology in step-up mode, including PCB inductance and an RC snubber circuit. Figure 6.24 displays the simulation waveforms of the converter in step-up mode when Vin is 30 V. As can be seen, the waveform in step-up mode is similar to that in step-down mode, with the only difference being a 180° phase shift. This indicates that the behavior of the topology in step-up mode mirrors its behavior in step-down mode.

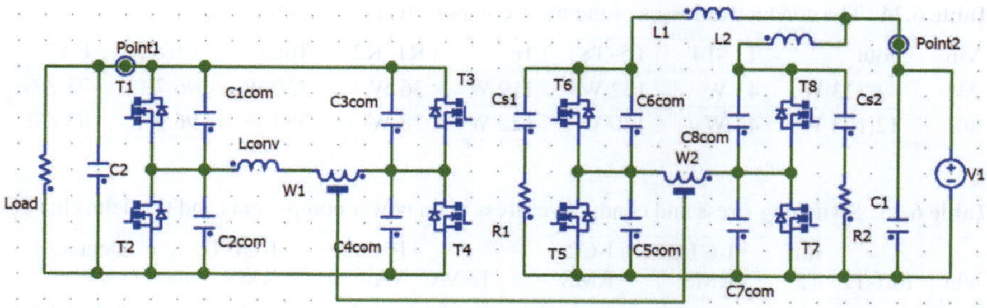

Fig. 6.23 The schematic of the DAB converter in step-up mode with the PCB inductance and RC snubber circuits

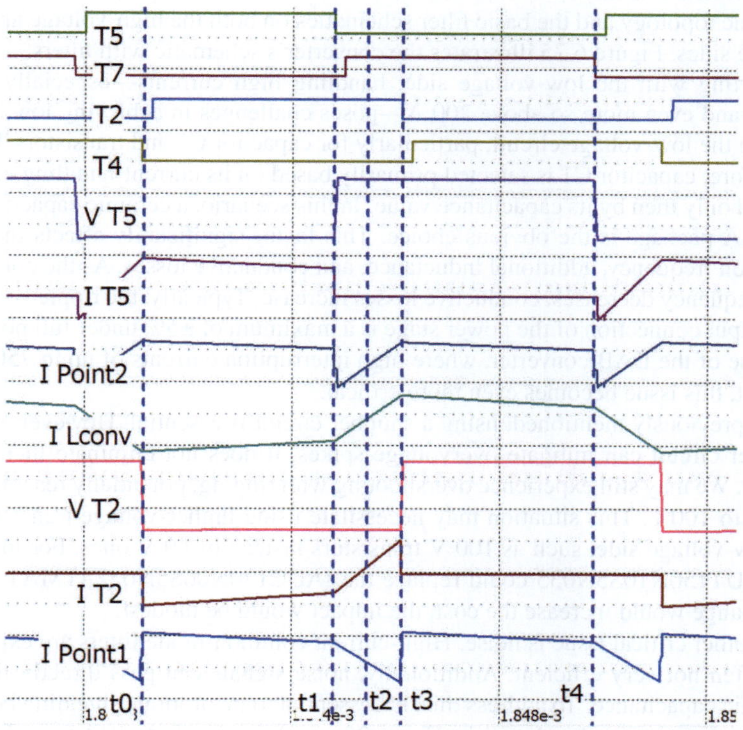

Fig. 6.24 The simulation waveforms of the dual-active bridge converter in step-up mode

Table 6.24 presents the conductive losses of the main components and the total losses when the input current reaches 400–430 A in step-up mode. Table 6.25 shows the switching stress, conductive stress on the main power components, and the delta changes in control required in step-up mode.

Table 6.24 The conductive losses of the main components and the total losses

Vin	Pout	T1–T4	T5–T8	Tr	R1–R2	Total	Eff	P-C
24	9723 W	43 W	132 W	119 W	36 W	330 W	96.7%	78.2%
30	12,104 W	44 W	140 W	122 W	78 W	384 W	96.9%	83.8%

Table 6.25 Switching stress and conductive stress main power components and the delta changes

Vin	Ioff-T2	Ioff T5	I-C1 RMS	I-C2 RMS	Tr-VA	Pout/Tr VA	I-LV Tr RMS	Degree/ ns/%
24 V	37 A	291 A	291 A	18.6 A	10,902	0.892	474 A	0.72/20/0.4
30 V	32 A	467 A	251 A	16 A	13,881	0.872	482 A	0.72/20/0.4

Now, let's consider the complete schematic of the DAB converter, which includes the basic topology and the basic filter schematics on both the high-voltage and low-voltage sides. Figure 6.25 illustrates the converter's schematic with filters.

Starting with the low-voltage side, handling high currents—especially above 100 A and even more so above 200 A—poses challenges in achieving low impedance in the low-voltage circuit, particularly for capacitor C1 and transistors T5–T8. Therefore, capacitor C1 is selected primarily based on its current-handling capability, and only then by its capacitance value. In this scenario, a ceramic capacitor with an SMA package is the obvious choice. This factor significantly affects the commutation frequency, additional inductance, and conductive losses. As the commutation frequency decreases, conductive losses increase. Typically, the ripple voltage at the output connection of the power stage is a maximum of ±5% under full power. In the case of the DAB converter, where high interruption currents of up to 750 A are present, this issue becomes even more critical.

As previously mentioned, using a snubber circuit is essential. However, while a snubber circuit can mitigate overvoltage spikes, it does not eliminate them completely. We may still experience overshooting with ringing, potentially reaching levels up to 100%. This situation may necessitate using higher-voltage transistors on the low-voltage side, such as 100 V transistors instead of 60 V ones. For instance, the IAUT150N10S5N035 could replace the IAUC120N06S5N032ATMA1. While this change would increase the cost, the impact would be modest.

Another critical issue is noise. High-current common-mode filters are expensive and often not very efficient. Additionally, noise signals can pass directly through parasitic capacitance. To address this, it's essential to incorporate shielding between the filtering stages. As shown in Fig. 6.25, we have implemented two stages of differential-mode filtering and two stages of common-mode filtering, with shielding in between.

The differential inductors (L3–L6) can be easily implemented using clamp cores, such as the 00K4119U090, on the busbar. This approach is practical since we need to use a busbar anyway to handle the 400 A current, which requires heavy conductive components. The capacitors C3 and C4 can be of any type.

Unfortunately, a high-voltage common-mode filter with good performance is prohibitively expensive. A better solution is to use a simple common-mode choke

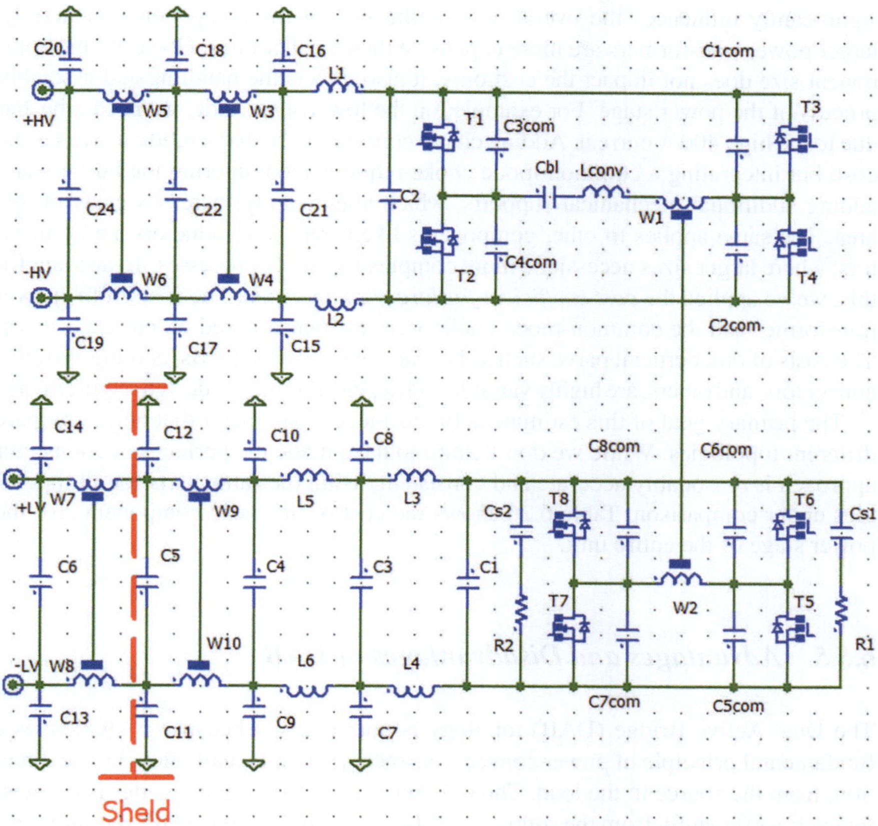

Fig. 6.25 The dual active bridge converter's schematic with filters

with two stages. This approach will result in better performance at a similar cost to a single high-performance common-mode choke. On the high-voltage side, it's also crucial to include a blocking capacitor (Cbl). When using an H-bridge, DC bias current will always be present, though the magnitude can vary. In the case of a DAB topology, there's a significant chance of encountering high-level transient DC bias currents. **The blocking capacitor is relatively small and inexpensive but is essential.**

6.5.4 Cost of the Entire Power Converter

Now we can make a rough estimate of the costs for the entire power stage. This estimate includes components available for purchase from an online vendor, with prices adjusted for production volumes of 1000 units per year. One critical factor that significantly influences the overall cost is the size of the components. Generally, larger power transformers are more expensive than smaller ones. However, the component size does not impact the cost only, it also affects the handling and assembly process of the power stage. For example, on the low-voltage side, we need a busbar due to the high 400 A current. Adding clamp cores to the busbar slightly increases the cost, but integrating a common-mode choke requires reconfiguring the busbars and adding additional mechanical supports, which significantly increases costs in this area. The same applies to other components like capacitors, inductors, and transistors, where larger sizes necessitate more complex assembly processes. To account for this, we've applied the *cost coefficient for larger components*. The costs of the power transformer and the common-mode choke were estimated based on our experience. The costs of other crucial parts, such as busbars, gate drivers, PCBs, cooling systems, connectors, and others, are highly variable and depend on specific design requirements.

The primary goal of this estimate is to introduce a parameter that helps compare different topologies. While we don't claim to have made the perfect selections, our approach is reasonably accurate and consistent, using the same parts for all topologies under comparison. Table 6.26 shows the cost of the basic components for the power stage of the entire unit.

6.5.5 Advantages and Disadvantages of DAB

The Dual Active Bridge (DAB) topology is controversial because it challenges a fundamental principle of power conversion: energy should ideally flow in one direction, from the source to the load. The ultimate goal of a power converter is to transfer energy efficiently from the source to the load by matching the load's requirements to the source's capabilities. This is where DAB differs significantly from other topologies, including its various modifications, which operate on the same basic idea.

In the DAB topology, energy flows in both directions—from the source to the converter and back, and from the converter to the load and back—all within a single half-cycle. While this does not contradict the laws of physics, it introduces several collateral effects. The apparent simplicity and broad applicability of DAB make it appealing, but in practice, it is not as straightforward as it seems. Design engineers often attempt to improve or correct the DAB topology, leading to numerous modifications and optimizations that, however, do not address the core principle. This suggests that much of the effort is directed at mitigating the consequences rather than addressing the root cause.

The primary disadvantage of the DAB topology is its inherent bidirectional energy flow, which leads to several other drawbacks: high levels of interrupting current, challenges in maintaining loop control stability, significant electromagnetic interference (EMI), and increased cost. As discussed in [49], the main power losses occur due to switching losses on the low-voltage side.

However, the DAB topology also has important advantages. It is inherently bidirectional, does not require synchronized rectification (since it does not operate in

Table 6.26 The cost of the basic components for the power stage of the entire unit

Parts	Part N	QTY	Price 1p $	Total $ parts	Total $ section
T1, 2, 3, 4	UF4SC120030K4S	4	19.96	79.84	
T5, 6, 7, 8	IAUT150N10S5N035	40	2.18	87.2	
W1–W2	Power Tr without holders	1	180.0	180.0	
C2	C4AQUBU4180A1XJ	4	0.99	3.96	
Lconv	E58/11/38	4	3.3	13.2	
Cbl	CGA6P1X7R1N106K250AC	8	0.29	2.32	
C1	CL32B106KBJNNNF	200	0.127	25.4	
Power supply for gate drivers T1–T4	AM2D-1218SZ	4	4.6	18.4	
Gate driver T1–T4	UCC53x0	4	0.91	3.64	
Power supply for gate drivers T5–T8	LMT78_15-1.0R	1	2.9	2.9	
Gate driver T5–T8	UCC5390-Q1	4	1.52	6.08	
Power stage	**Total**	**1**			**422.94**
C1com–C4com	R76UI056050H4J	8	0.2	1.6	
C5com–C8com	CKC18C473MCGAC7210	80	0.98	78.4	
Cs1–Cs2	C1210C125M1RAC7800	16	0.16	2.56	
R1–R2	PWR163S-25-2R00FE	20	1.68	33.6	
Aux circuits	**Total**	**1**			**116.16**
L1, 2	CPEX3231A-270MC	2	11.8	23.6	
W3, 4, 5, 6	B82726S6243A040	2	9.4	18.8	
C3, 4, 5	C4AQUBU4180A1XJ	5	0.99	4.95	
L3, 4, 5, 6	00K4119U090	8	1.5	12.0	
C6, 7, 8, 9	CL32B106KBJNNNF EMHS500ARA361MKE0S	30 6	0.127 0.65	3.81 3.9	
W7–W10	High current common mode choke	2	60	120.0	
Filters	**Total**	**1**			**187.06**
Summary	Power stage	Aux	Filters		739.36
Coefficient	On big size of parts	1.5			369.68
	Total cost of converter				**1109.04**

synchronized rectification mode), and offers higher efficiency in the power transformer due to the trapezoidal waveform of the current passing through it, along with a higher utilization factor of the transformer.

6.5.5.1 Operation and Application Area of DAB

Now that we've discussed the advantages and disadvantages of the DAB topology, let's explore its potential application areas. Based on the analysis, it's evident that the basic DAB topology is not suitable for applications involving high currents, such as 400 A. The primary application area for DAB is in bidirectional converters with a maximum power of around 2 kW and limited current, where cost-effectiveness is also a consideration. Table 6.27 outlines these preferred application areas. It's important to note that this information should not be taken as absolute; it is flexible depending on the specific goals of the application.

6.5.5.2 Modifications

The various disadvantages of DAB converter have resulted in numerous attempts to address its shortcomings. As mentioned earlier, DAB holds the record for the number of modifications for a DC–DC converter, and some of these are particularly noteworthy. For example, in [44], a modification involves the use of three power transformers and 12 active switches. However, the practicality of such an expensive unit for transportation applications is highly questionable.

Another significant modification is the development of the next generation of DAB topology, known as the Dual Active Half Bridge (DAHB). Also referred to as the asymmetrical DAB, the DAHB retains the same fundamental principles as the DAB and, consequently, shares many of the same disadvantages.

The main difference is that the DAHB topology places the inductor on the low-voltage side, which reduces the issue of high ripple current on that side. Additionally, it only uses four active switches instead of eight.

The authors of [50] describe the DAHB as "A natural ZVS medium-power bidirectional DC–DC converter with a minimum number of devices," but a closer look at the oscillograms in Fig. 6.21 of the article reveals hard switching with significant voltage overshoot and high levels of EMI. Despite its apparent simplicity, the DAHB has attracted the interest of design engineers who have attempted to refine and optimize it [51] and in many other directions. However, it's important to recognize that DAHB is a deep modification of DAB, rather than an entirely new topology.

Table 6.27 Possible areas of operation and application for the basic DAB converter

Power conversion	78–91%	
Co/XMFR	70–90%	
Stab/coeff	0.4 step-up	0.6 step-down
Max commutation frequency	100 kHz	Limited by switching current, cost
Start-up method	Complicated	Problem is hard commutation
Reliability	Low	All protection by control
Input voltage max	750 VDC	Limited by semiconductors 1200 V
Input voltage min	30–100 VDC	Below 100 VDC increase cost
Input current	Max 100 A	Limited by switching current, cost
Range of Vin	1.3 times	Vout is fixed
Load range	60–100%	With soft-switch this range
Output current	Max 100 A	Limited by switching current, cost
Response time	Slow	Limit by Lconv and stability coef.
Directional	Step-up or step-down	Straight bi-directional

6.6 Resonance with PWM Topology

In 1998, the authors were tasked with developing a step-up isolated DC–DC converter with the following specifications: an input voltage range of 10–16 VDC (with the option to extend to 8–18 VDC), an output voltage of 400 VDC, and an output power of up to 4.5 kW for an automotive application. This requirement meant that the converter needed to be both cost-effective and reliable. To achieve this, it was crucial to minimize high-current commutation on the low-voltage side, favoring a sinusoidal-resonant operation, while focusing the control primarily on the high-voltage side, where current levels are significantly lower. At the time, the existing topologies, such as Phase-shift, Dual Active Bridge (DAB), and LLC resonant converters, were not suitable for this application due to their high commutation currents and their inability to efficiently handle a wide input voltage range—at least 1.6 times the minimum value. These limitations would have led to a significant increase in cost, which was absolutely unacceptable for automotive applications. This challenge motivated the authors to develop a new topology specifically designed to meet the demanding requirements of this use case.

6.6.1 Description of Resonance with PWM Converter

Figure 6.26 shows the basic schematic of the Resonance with PWM Converter. The first publication about this topology was in 2001 [52], and it was patented in 2002 [53]. In 2008, a control algorithm for this topology was also patented [54]. This topology can transfer energy in both directions without any modifications [52, 55–57]. Further information about this topology and its applications can be found in other publications from the last several years [58–62]. One important feature of this topology is that it does not have modifications that significantly alter its fundamental behavior. It belongs to the family of series resonance topologies with a clamp diode, where regulation is performed using Pulse Width Modulation (PWM).

The topology with a clamp diode is sometimes referred to as a “dose energy transfer” topology. This means that in the resonance topology with a clamp diode, the system can automatically change its mode from a voltage source to a power

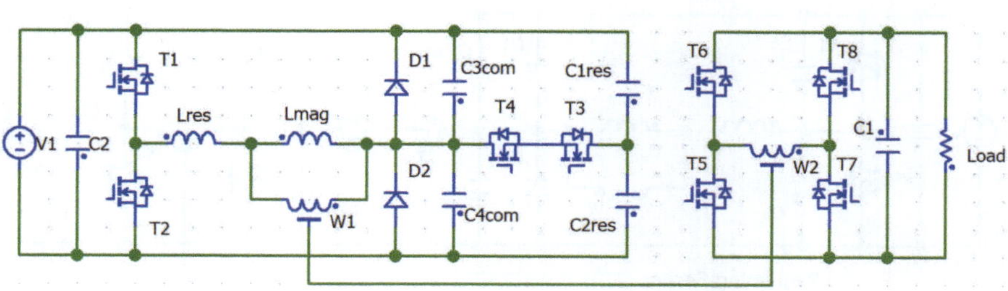

Fig. 6.26 The schematic of the basic Resonance with PWM Converter

source as the load increases, without the need for external control. In this mode, during an overload condition, the energy source will not experience an overcurrent, effectively protecting the system from short-circuit conditions at the output. This gives the topology a higher level of reliability compared to topologies that require active control during overload conditions.

Figure 6.27 illustrates the basic topology with clamp diodes. The resonance capacitor can be calculated using the following equation:

$$C_{\text{RES}} = \frac{P\text{out}}{f\text{com} \times \text{Vin}^2} \tag{6.9}$$

where: $C_{\text{RES}} = C_{1\text{res}} + C_{2\text{res}}$ and f_{COM} is the maximum commutation frequency.

Figure 6.28 shows the schematic used for simulation. The regulation is achieved through PWM control, but this topology can also be controlled using a combination of frequency modulation and PWM [54]. The regulating section is bidirectional, meaning it can be placed on either side of the transformer—on the primary or the secondary side. This flexibility allows for power conversion in both directions. However, it's important to note that while this topology supports bidirectional power flow, it is not fully bidirectional in all aspects. In other words, the topology can function in either step-up or step-down mode, but the output voltage will differ based on the transformer turns ratio, or it may require a change in the transformer's ratio for different voltage levels.

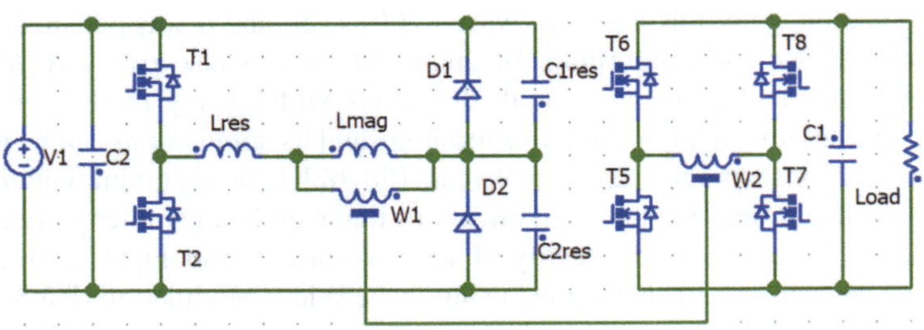

Fig. 6.27 The basic topology with clamp diodes

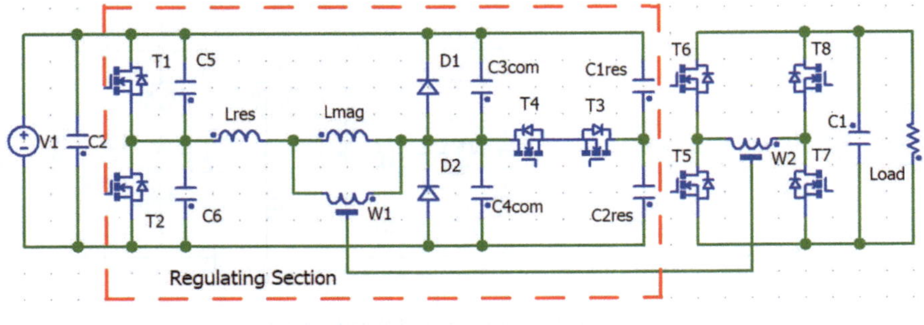

Fig. 6.28 The schematic used for simulation

6.6.2 Operation of Resonance Topology with PWM

Let's consider the operation of the Resonance with PWM topology in step-down mode under the following conditions:

- **Input voltage (V1)**: 700 V
- **Output voltage (Vload)**: 30 V
- **Maximum power**: 12 kW.
- **Commutation frequency**: 100 kHz
- **Resonance inductor (Lres)**: 5.65 μH
- **Magnetizing inductor (Lmag)**: 55 μH
- **Resonance capacitors (C1res and C2res)**: 198 nF each
- **Commutation capacitors (C3com and C4com)**: 10 nF each
- **The resonance frequency** is 104 kHz
- **Additional capacitors C5 and C6** are each 0.4 nF
- **Power switches T1 and T2**: two UF4SC120030K4S transistors connected in parallel
- **Power switches T3 and T4**: are UJ4SC075011K4S
- **Power switches T5–T8**: 10 IAUC120N06S5N032ATMA1 transistors connected in parallel
- **Turn ratio of power transformer**: 9

6.6.2.1 Design Considerations

The transformer turn ratio is calculated using the following equation:

$$N = {}^{Vinmin}\!\!/_{2 \times Vout} \tag{6.10}$$

$$N = {}^{550}\!/_{2 \times 30} = 9.16$$

This results in a transformer ratio of 1:9. All simulations were conducted under non-optimal conditions to demonstrate the basic implementation and the potential of the topology. The commutation frequency was selected based on the influence of the PCB on the low-voltage side and the additional losses in the power transformer. Operating at a frequency above 100 kHz is challenging for this current level. Moreover, this affects the cost, unfortunately leading to an increase. The efficiency drops by 1% when the frequency reaches 150 kHz due to increased PCB losses on the low-voltage side or the need for a PCB with lower losses, which would further increase the cost. Figure 6.29 displays the simulation waveforms of voltage and current of the converter at maximum commutation frequency with regulation performed by PWM. Figure 6.30a–g illustrate the equivalent circuits for each operating mode.

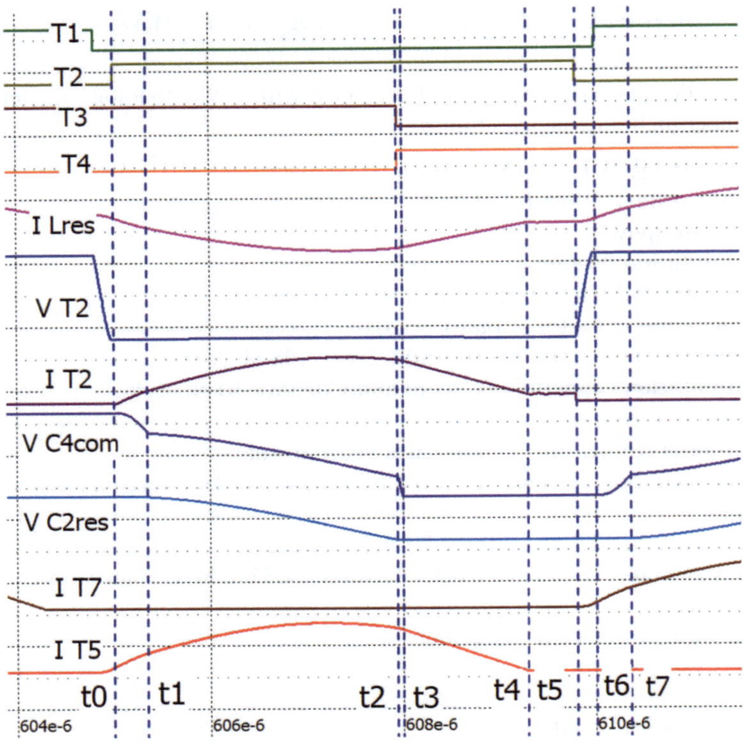

Fig. 6.29 The simulation waveforms of voltage and current of the converter at maximum commutation frequency with regulation performed by PWM

6.6.3 *Description of Operation*

Starting with Fig. 6.30a, which shows the time segment t0–t1, the device T2 turns on and begins conducting. The current starts at zero and rises sinusoidally. The value and waveform of this current are determined by the resonant circuit (C3com, C4com, and Lres) and the load value. The devices T5 and T8 begin conducting at the commutation time ZCS and continue operating during this period. T2, C3com, and C4com are active during the time interval t0–t1.

At time t1, the voltage across C4com and C2res becomes equal, and devices T3 and T4 begin to conduct, as shown in Fig. 6.30b. They start operating under pure ZVS conditions. At this point, C1res and C2res also become active. The current through T2, T3, T4, T5, and T8 continues to rise sinusoidally. The value and waveform of this current are determined by the combined resonant circuit (C3com + C4com + C1res + C2res) and Lres, as well as the value of the load.

At t2, switch T3 turns off, interrupting the current, and simultaneously, T4 turns on, preparing for the next half-cycle. This interruption occurs under ZV conditions because T3 and T4 are positioned between the capacitors C1res and C2res on one

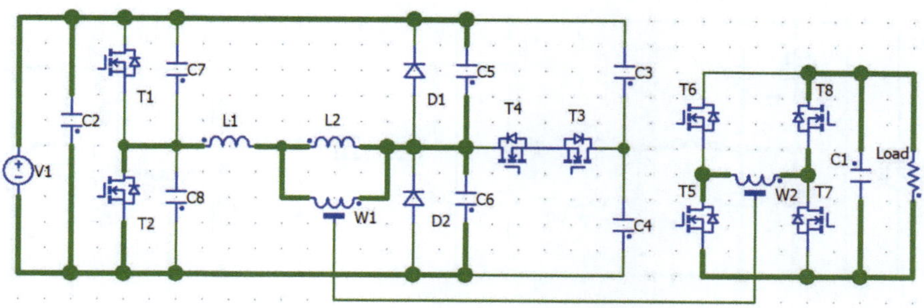

Fig. 6.30 (**a**) Equivalent circuit of the Resonance with PWM topology between t0 and t1

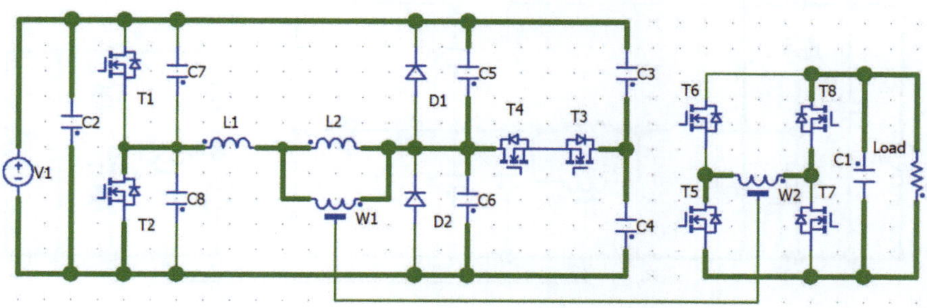

(**b**) Equivalent circuit of the Resonance with PWM topology between t1 and t2

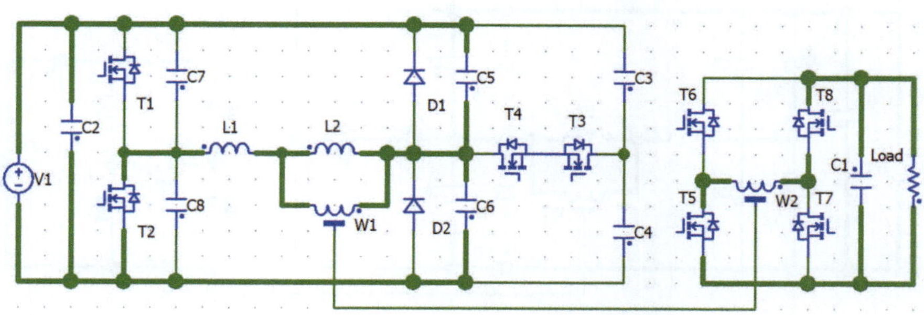

(**c**) Equivalent circuit of the Resonance with PWM topology between t2 and t3

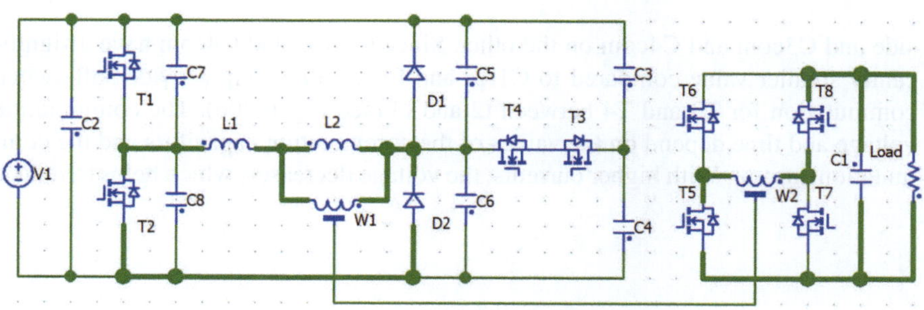

(**d**) Equivalent circuit of the Resonance with PWM topology between t3 and t4

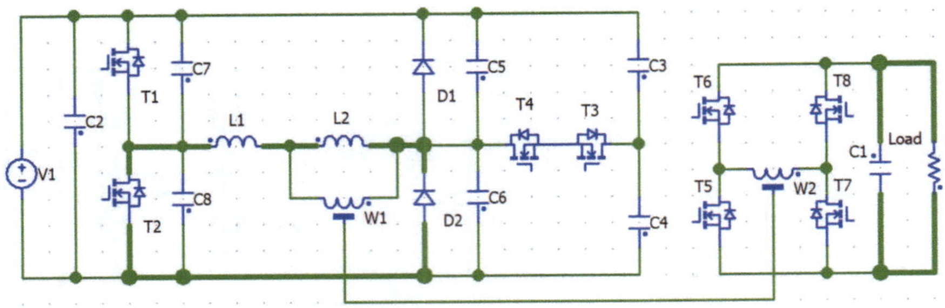

(**e**) Equivalent circuit of the Resonance with PWM topology between t4 and t5

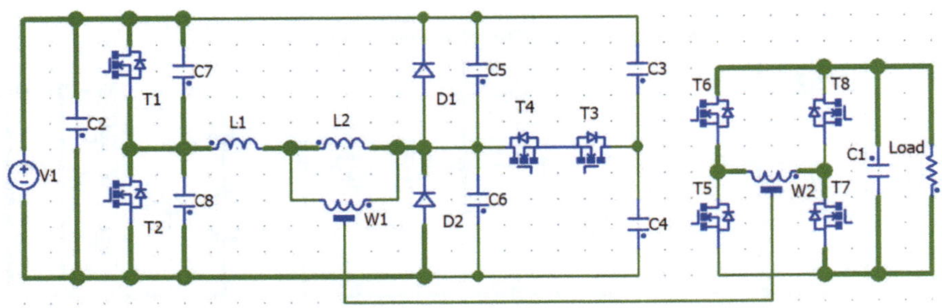

(**f**) Equivalent circuit of the Resonance with PWM topology between t5 and t6

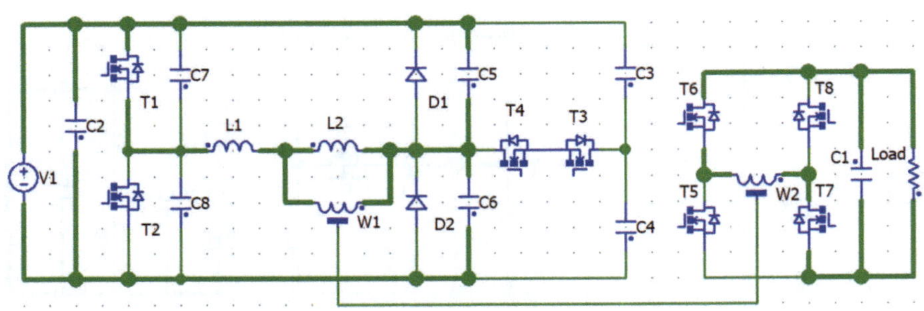

(**g**) Equivalent circuit of the Resonance with PWM topology between t6 and t7

side and C3com and C4com on the other. Since C3com and C4com have a significantly smaller value compared to C1res and C2res, they help provide soft-switch commutation for T3 and T4 between t2 and t3 (see Fig. 6.30c). The commutation voltage and time depend on the values of the commutation capacitors and the commutation current. With higher currents, the voltage decreases, which helps to reduce the EMI level during the period from t2 to t3. Also, at t2, the transfer of current/energy from the source to the load stops, and this process happens relatively quickly.

At t3, the energy discharge process of the inductor Lres begins. Between t3 and t4, current flows in the circuit through D2, Lres, Tr, and T2 (Fig. 6.30d), linearly decreasing to zero.

At t4, the current flowing through the low-voltage side in T5 and T8 decays to zero, and the power conversion process, transferring energy to the load, stops. Between t4 and t5 on the high-voltage side, the magnetizing current flows via the circuit comprising T2, Tr, Lres, and D2 (Fig. 6.30e).

At5, switch T2 turns off, interrupting the magnetizing current, which has a relatively small value. The body capacitance of T2 and the additional capacitors C5 and C6 are sufficient to provide ZV turn-off. Between t5 and t6 (Fig. 6.30f), the magnetizing current charges the body capacitance of T1, T2, and C5, C6, preparing T1 for ZVS turn-on.

At t6, T1 turns on, and a new half-cycle of power conversion begins. This new half-cycle is a mirror image of the half-cycle between t0 and t6. Between t6 and t7, current flows through T1, Tr, Lr, and C3com–C4com (Fig. 6.30g). This explanation demonstrates the operation of the Resonance with PWM topology in a step-down mode and highlights its ability to handle high power levels while maintaining efficient power conversion.

6.6.3.1 Simulation Results Analysis

We can now consider the results of the simulations. This topology is completely soft-switching, so for calculating losses, we used only the conductive losses of the main power components. In real-life applications, it is necessary to include the conductive losses from the PCB, busbars, and filters. These losses can reach up to 1%, depending on cost requirements and the cooling methods employed. Table 6.28 shows the conductive losses of the main components and the total losses at maximum output power. As you can see, the main losses occur in the output rectifier and power transformer. The parameter "*P–C*" represents the duty cycle of power conversion.

Table 6.29 presents the switching stress and conductive stress on the main power components. From this data, we observe that the using factor of the power transformer is around 90%.

Table 6.28 Conductive losses of main components and total losses step down of Resonance topology with PWM

Vin	Pout	T1–T2	T3–T4	Tr	T5–T8	Total	Eff	P-C
550 V	12,484 W	60 W	92 W	130 W	131 W	415 W	96.7%	96.0%
700 V	11,931 W	59 W	77 W	128 W	131 W	395 W	96.7%	87.4%
800 V	12,000 W	61 W	73 W	135 W	141 W	410 W	96.7%	85.4%

Now, let's consider the simulation results under different operating conditions, such as varying input voltage and load levels. We aim to evaluate the converter's performance and understand what happens when the input voltage increases. Table 6.30 presents the main performance characteristics of the converter with a higher input voltage. We observe a significant increase in stress on the commutation capacitors, both in terms of current and voltage. Additionally, at higher voltages, variable frequency control is required when operating under low load conditions [3]. This means that the topology can handle situations where the input voltage varies by up to three times.

Table 6.31 shows the results of simulations at different power levels and input voltages. This simulation was conducted with an additional commutation inductor (Lcom, see Fig. 6.8), which helps to linearize the commutation/switching current of T1 and T2. Now, the question arises about which mode of power conversion should be considered under abnormal conditions, specifically in worst-case scenarios, such as when there is an overload of at least twice the normal load without current protection or current limits. Evaluating these conditions helps estimate the reliability of the power stage itself. Of course, protection must be implemented, but it requires a response time.

Table 6.32 presents the results under overload conditions. As we can see, in this situation, the clamp diodes D1 and D2 become very active. This table serves as a reference for selecting the appropriate power components. It is also important to

Table 6.29 Switching stress and conductive stress main power components

Vin	Ioff-T2	Ioff T3	Voff T3	t2–t3	I-C2res	I-C2 RMS	I-C1 RMS	Tr-VA	Pout/Tr VA	I-LV Tr
550 V	13.7 A	0	0	0	25.3 A	13.7 A	200 A	13,560	0.92	452 A
700 V	14.0 A	71 A	158 V	46 ns	21.2 A	16.2 A	235 A	13,572	0.88	445 A
800 V	13.5 A	83 A	240 V	58 ns	21.4 A	16.9 A	287 A	13,717	0.875	483 A

Table 6.30 Main performance of converter higher level Vin

Vin	Pout	Tr-VA	Pout/TrVA	I-Ccom	Eff	P-C
800 V	12,000 W	13,717	0.875	5 A	96.7%	85.4%
1100 V	12,241 W	14,330	0.854	7.75 A	96.5%	75%
1650 V	12,160 W	14,541	0.836	11.5 A	96.4%	68.6%

Table 6.31 Different power level with different input voltage

Vin	Pout 6000 W		Pout 1200 W	
	Ioff T3	Eff	Ioff T3	Eff
550 V	12.3 A	97.4%	7.6 A	98.0%
700 V	11.2 A	96.9%	7.8 A	98.9%
800 V	11.6 A	96.5%	9.1 A	96.9%

Table 6.32 Result under overload conditions

Vin	Load ohm	Pout	Eff	I-T2 RMS	I-LV Tr	I-D1/D2 Peak	I-D1/D2 average
550 V	0.0375	11,316 W	93.5%	49.3 A	600 A	78 A	7.9 A
800 V	0.0375	23,364 W	94.0%	66.5 A	867 A	120 A	13.4 A
800 V	0.025	21,160 W	91.8%	78.2 A	997 A	137 A	18 A

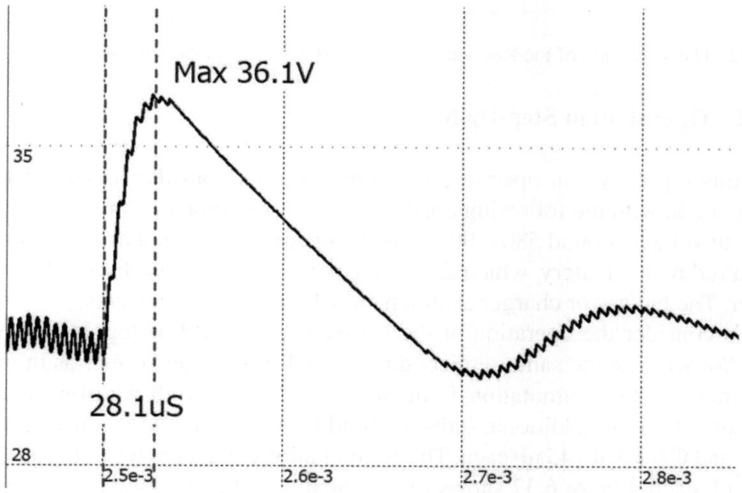

Fig. 6.31 The simulation of output voltage and timing when change level of load 10 times

Table 6.33 The delta changes in control to change Vout on 0.3 V when the output power is 12 kW

Vin	600 V	700 V	800 V	Average
Degree	2.52°	1.48°	1.26°	1.75°
Switching time (ns)	70 ns	41 ns	35 ns	48.7 ns
Control percentage	1.4%	0.82%	0.70%	0.97%

note that when Vin is at 550 V, the power stage operates completely in dose energy mode.

Another abnormal condition to consider is a rapid disconnection of the full load, also known as load dump mode. Figure 6.31 shows the simulation of this condition when the load changes drastically, from the full load of 12 kW/0.075 ohm to 1.2 kW/0.75 ohm. The capacitor C1 is 1000 μF, without any additional filters, representing the worst-case scenario. We observe a 6 V overshoot, but with full filtering and a battery, the overshoot would be reduced to a maximum of 0.7 V. Table 6.33 shows the delta changes in control required to change Vout by 0.3 V when the output power is 12 kW. The average control adjustment needed is 1.75°.

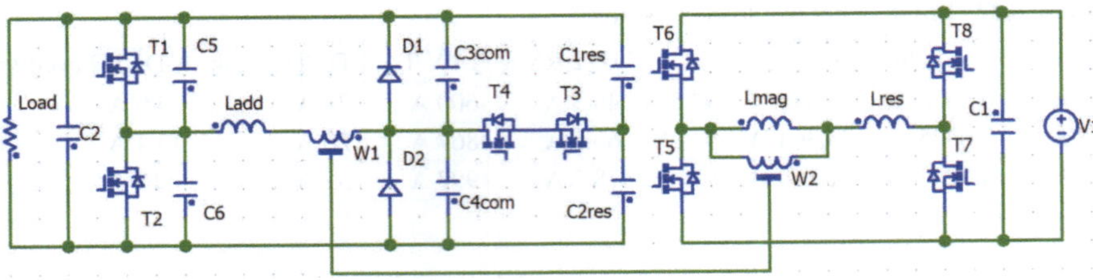

Fig. 6.32 The schematic of the Resonance with PWM topology in step-up mode

6.6.3.2 Operation in Step-Up Mode

Since this topology can operate bidirectionally, let's consider its performance in step-up mode with the following conditions: an input voltage (Vin) of 24–30 V and an output voltage around 580–600 V. The input current must not exceed 400–430 A, as required by the battery, which acts as the power source or is charged by a special charger. The battery or charger cannot handle higher output currents.

Let's consider the operation of the Resonance with PWM topology in step-up mode. We will use the same semiconductors and power transformer as in the step-down mode. The commutation frequency is 100 kHz, with a resonant inductor (Lres) of 30 nH, an additional inductor (Ladd) of 2.7 μH, and resonant capacitors (C1res and C2res) of 145 nF each. The commutation capacitors (C3com and C4com) are 10 nF each. Figure 6.32 shows the schematic of the topology in step-up mode, while Fig. 6.33 illustrates the simulation waveforms of voltage and current of the converter in this mode when the input voltage (Vin) is 30 V, and the output voltage is 588 V. In this operation mode, devices T1and T2 function primarily through their body diodes.

Table 6.34 presents the conductive losses of the main components and the total losses when operating at maximum output power in step-up mode. As you can see, the behavior of the regulating section in step-up mode is similar to that in step-down mode. The authors have used this topology in both modes, achieving a current of 475 ADC on the low-voltage side, with an overall efficiency of 94.4% for the entire unit under full power conditions. The losses accounted for include the PCB, power connections, busbars, filters, and other components.

6.6.3.3 Whole Schematic of the Resonance with PWM Converter

Now, let's consider the complete schematic of the converter, which includes both the basic topology and the filter circuits on both the high-voltage and low-voltage sides. Figure 6.34 shows the overall schematic of the converter with filters.

Starting with the low-voltage side: the challenge with high currents—particularly when they exceed 100 A, and even more so when they exceed 200 A—is the difficulty in implementing a low-impedance circuit for capacitor C1. Therefore, capacitor C1 is primarily selected based on its current-handling capability, followed

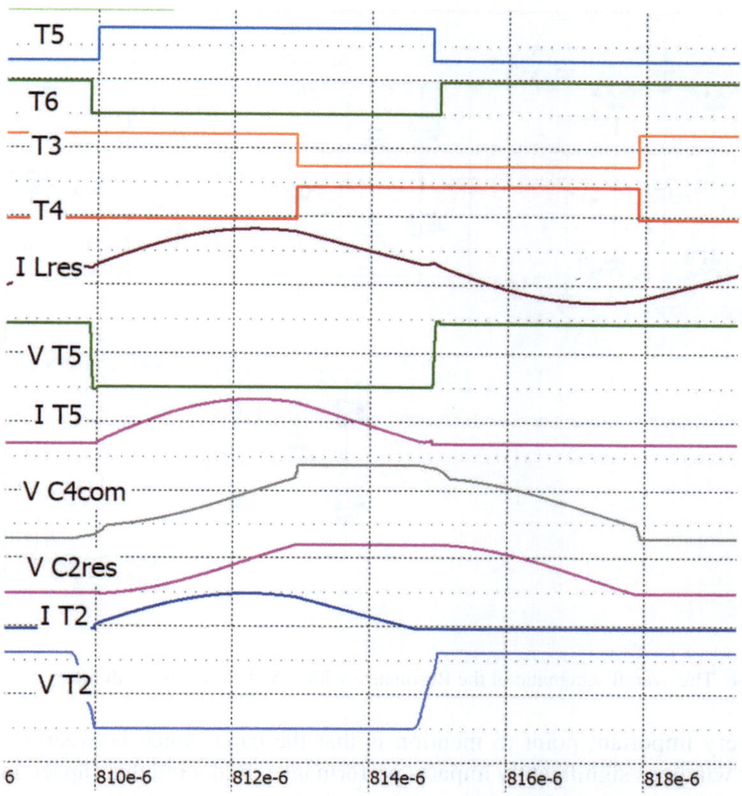

Fig. 6.33 The simulation waveforms of voltage and current of the converter in step down mode

Table 6.34 Conductive losses of main components and total losses in step-up

Vin	Pout	T5–T8	T1–T2	T3–T4	Tr	Total	Eff	P-C
24 V	9949 W	143 W	45 W	65 W	127 W	380 W	96.2%	95.4%
30 V	12,160 W	136 W	44 W	51 W	122 W	353 W	97.1%	93.4%

by its capacitance value. In this situation, it is crucial to use ceramic capacitors in an SMA package. This choice significantly affects the commutation frequency, additional inductance, and conductive losses. Typically, the ripple voltage on the output connection of the power stage is a maximum of ±5% under full power conditions.

The differential inductors L3–L6 can be implemented relatively easily by using a clamp core, such as the 00K4119U090, on the busbar, which is necessary anyway due to the high current of 400 A, requiring robust conductive components. Capacitors C3 and C4 can be of any suitable type for this application. It's worth noting that a common-mode filter is not required on the low-voltage side, as implementing one would be both complex and costly. This is made possible by the Zero-Current Switching (ZCS) and Zero-Voltage Switching (ZVS) on the low-voltage side, which result in low switch amplitude and a low dV/dt.

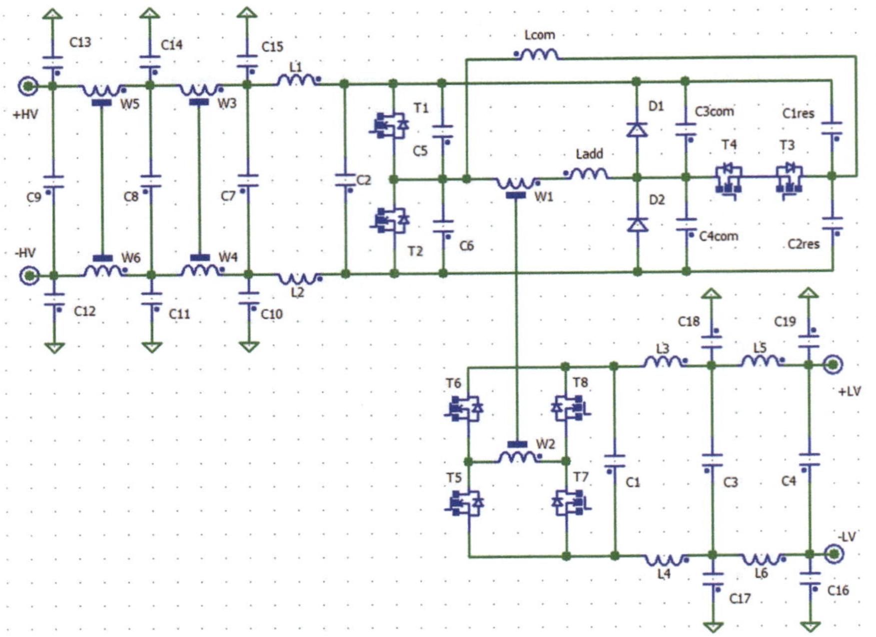

Fig. 6.34 The overall schematic of the Resonance with PWM Converter with filters

A very important point to mention is that the capacitance between the transformer windings significantly impacts performance. Reducing this capacitance by a factor of five can help lower electromagnetic interference (EMI) by as much as 20 dB. Additionally, the way the power transformer is connected to the power stage is crucial. The authors have consistently used transformers with winding capacitance values below 100 pF, preferably in the range of 50–70 pF. This is the type of transformer considered in these calculations, effectively ignoring the impact of inter-winding capacitance.

On the high-voltage side, the situation is more challenging but still manageable. The high amplitude and relatively high dV/dt values of up to 6–8 kV/μs require a common-mode filter. Unfortunately, high-voltage common-mode filters with good performance tend to be very expensive. In this case, a better approach is to use a simpler two-stage choke. This solution provides better performance at a similar cost to a single high-quality choke.

6.6.4 Cost Estimation of the Converter

Now we can estimate the cost of the entire power stage, including both the power stage itself and the filters on the low-voltage and high-voltage sides. The costs of critical components, such as busbars, gate drivers, PCBs, cooling systems, connections, and many others, can vary significantly. The main objective of this estimation is to introduce a parameter that can help facilitate comparisons between different topologies: the coefficient related to the size of the components. Of course, we do not claim that we have made the best selections, but we believe they are sufficient for this analysis. We will use the same components for evaluating other topologies as well.

Table 6.35 presents the costs of the basic components for the power stage of the entire unit. It is worth noting that the transistors T1 and T2 can be replaced with diodes D3 and D4 when the topology is intended for use in a step-up mode with regulation occurring after the transformer. In this case, the cost of the converter decreases.

Table 6.36 shows the costs of the basic components for the power stage of the entire unit specifically for the step-up mode. Of course, we do not pretend that we

Table 6.35 Cost of basic components for the power stage of the whole unit in step-down mode

Parts	Part N	QTY	Price 1p $	Total $	Total $
T1, 2	UF4SC120030K4S	4	19.96	79.84	
T3, 4	UJ4SC075011K4S	2	23.2	46.4	
T5, 6, 7, 8	IAUC120N06S5N032	40	0.73	29.2	
D1, 2	S4D30120H	2	6.74	13.48	
W1–W2	Power Tr without holders	1	150.0	150.0	
Ladd	E58/11/38	4	3.3	13.2	
C1res, C2res	R76UN22204040J	18	0.63	11.34	
C2	C4AQUBU4180A1XJ	3	0.99	2.97	
C1	CL32B106KBJNNNF	130	0.127	16.51	
Power supply for gate drivers T1–T4	AM2D-1218SZ	3	4.6	13.8	
Gate driver T1–T4	UCC53x0	4	0.91	3.64	
Power supply for gate drivers T5–T8	LMT78_15-1.0R	1	2.9	2.9	
Gate driver T5–T8	UCC5390-Q1	4	1.52	6.08	
Power stage	Total	1			**389.36**
C3com, C4com	R76UI1470SE40J	4	0.28	1.12	
Lcom	PFC3819QM-151K11B-00	1	11.4	11.4	
Aux circuits	Total	1			**12.52**
L1, 2	CPEX3231A-270MC	2	11.8	23.6	
W3, 4, 5, 6	B82726S6243A040	2	9.4	18.8	
C7, 8, 9	C4AQUBU4180A1XJ	4	0.99	3.96	
L3, 4, 5, 6	00K4119U090	8	1.5	12.0	
C3, 4	CL32B106KBJNNNF EMHS500ARA361MKE0S	30 6	0.127 0.65	3.81 3.9	
Filters	Total	1			**66.07**
Power stage	Aux circuits, filters				467.95
Coefficient	On big size of parts	1.2			93.59
	Total cost of converter				**561.54**

Table 6.36 Cost of basic components for the power stage of the whole unit when the unit is intended for use only in step-up mode

Parts	Part N	QTY	Price 1p $	Total $	Total $
D3, 4	S4D30120H	2	6.74	13.48	
T3, 4	UJ4SC075011K4S	2	23.2	46.4	
T5, 6, 7, 8	IAUC120N06S5N032	40	0.73	29.2	
D1, 2	S4D30120H	2	6.74	13.48	
W1–W2	Power Tr without holders	1	150.0	150.0	
Ladd	E58/11/38	4	3.3	13.2	
C1res, C2res	R76UN22204040J	18	0.63	11.34	
C2	C4AQUBU4180A1XJ	3	0.99	2.97	
C1	CL32B106KBJNNNF	130	0.127	16.51	
Power supply for gate drivers T3–T4	AM2D-1218SZ	1	4.6	4.6	
Gate driver T3–T4	UCC53x0	2	0.91	1.82	
Power supply for gate drivers T5–T8	LMT78_15-1.0R	1	2.9	2.9	
Gate driver T5–T8	UCC5390-Q1	4	1.52	6.08	
Power stage	**Total**	1			**312.04**
C3com, C4com	R76UI1470SE40J	4	0.28	1.12	
Lcom	PFC3819QM-151K11B-00	1	11.4	11.4	
Aux circuits	**Total**	1			**12.52**
L1, 2	CPEX3231A-270MC	2	11.8	23.6	
W3, 4, 5, 6	B82726S6243A040	2	9.4	18.8	
C7, 8, 9	C4AQUBU4180A1XJ	4	0.99	3.96	
L3, 4, 5, 6	00K4119U090	8	1.5	12.0	
C3, 4	CL32B106KBJNNNF EMHS500ARA361MKE0S	30 6	0.127 0.65	3.81 3.9	
Filters	**Total**	1			**66.07**
Power stage	Aus circuits, filters				390.63
Coefficient	On big size of parts	1.2			78.13
	Total cost of converter				**468.76**

made the best selections, but more or less, they are good enough. For other topologies, we will use the same parts.

6.6.5 Advantages and Disadvantages of Resonance with PWM Topology

Now we can see that this topology is workable, but of course, it has disadvantages and advantages. Let us consider this point and start, of course, with disadvantages, because if we overlook them, sooner or later, they will show problems. Also, maximum attention should be given to this aspect, but please note that any way during the implementation process, there are always some issues. **Advantages do not highlight problems; they don't have this habit.** By the way, we, the authors, strongly recommend using this order: first consider disadvantages and then advantages.

The main disadvantage of this topology is the AC switch, T3 and T4. The AC switch consists of two series-connected devices, resulting in twice the conductive losses. In this case, it is necessary to select transistors with minimum R_{DS_ON} and diodes with minimal forward voltage drop. Additionally, we can use synchronous rectification, which helps but complicates control. Another problem with the AC switch is that all inductors of the bond wire of transistors and the inductor of the connection between devices are connected in series. Figure 6.35 shows a schematic of these connections. In this case, we need to be prepared for a total inductance of around 20 nH or possibly more. Figure 6.36 shows the equivalent schematic of these connections. When the transistor is fast turned off, we get a ringing/resonance circuit. This can cause overvoltage and excessive energy loss, which leads to premature failure, especially when using WBG transistors, which are faster than Si MOSFETs/IGBTs.

In this topology, the AC switch turn-off is pure ZVS (Zero Voltage Switching), so increasing the gate resistor increases the internal turn-off time. Unfortunately, this is another point that limits the increase in commutation frequency and the

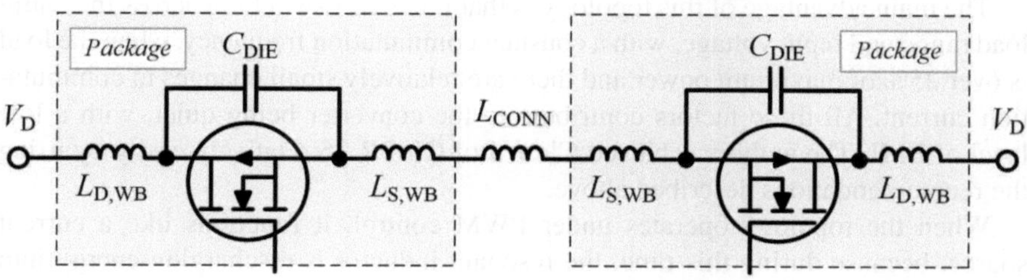

Fig. 6.35 The schematic of connections two transistors in series to form a bidirectional switch

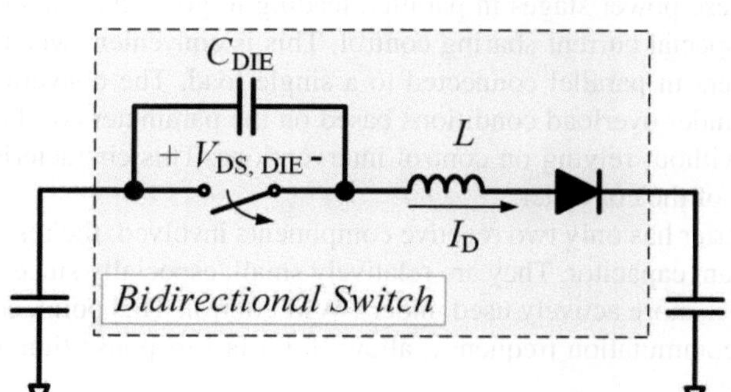

Fig. 6.36 The equivalent schematic of connections two transistors in series

maximum interruption current through this switch. All these factors indicate that this disadvantage demands serious attention. Fortunately, the transistors, T3 and T4, for these switches can be selected with voltages that are two times smaller than those for switches T1 and T2.

Next, there are significant changes in commutation current, approximately three times, when the regulating section is before the transformer. This problem depends on the range of input voltage and changes in load. For this issue, attention must be paid to the magnetizing inductor and gate resistor for turning on T1 and T2. Alternatively, it is possible to use an additional inductor of around 200–400 μH (see Fig. 6.34), which resolves the problem entirely.

Another disadvantage is providing a soft start for the converter. When using variable frequency control, this is not a problem, but when only constant frequency control is used, it is preferable to pre-charge the output capacitor when the regulating section is after the transformer. In this mode, it is also important to ensure that the power transformer does not exceed its maximum voltage rating. In other words, the maximum input voltage is what should be used for calculating the transformer. This requirement leads to an increase in the size of the power transformer.

Another negative point, applicable to all types of resonance converters, is the relatively high peak current through semiconductors and the windings of the transformer. The consequence of this disadvantage is an increase in the VA rating of the power transformer. The authors are confident that this disadvantage is overshadowed by the advantages of this topology.

The main advantage of this topology is that it has soft switching across the entire load range and input voltage, with a constant commutation frequency, when the load is over 25% of maximum power and there are relatively small changes in commutation current. All these factors contribute to the converter being quiet, with a low level of EMI. The authors achieved Class 4 of CISPR 25 relatively easily by using the recommendations described above.

When the topology operates under PWM control, it functions like a current source, because during this time, the resonant inductor is discharging energy into the load via the transformer (time period t3–t4, Fig. 6.4d). This high impedance condition and strong control through energy packets facilitate the connection of two or more of these power stages in parallel, leading to good current sharing without the need for special current-sharing control. This is convenient when using two or more converters in parallel connected to a single load. The converter has limited input power under overload conditions based on the parameters of the components themselves, without relying on control interventions. This characteristic increases the reliability of the converter.

This converter has only two reactive components involved: the resonant inductor and the resonant capacitor. They are relatively small, especially since only the resonant inductor is more actively used under PWM control. This point, combined with the constant commutation frequency, allows for a fast response time in controlling this topology.

Table 6.37 Possibility of operation and application areas of basic Resonance topology with PWM, regulation is after transformer

Power conversion	85–96%	
Co/XMFR	87–92%	
Stab/coeff	0.8, 0.9, 1.17	0.957 average
Max frequency	1000 kHz	Limited by passive power components
Start-up method	Itself	
Reliability	High	Dose energy capacitors
Input voltage max	850 VDC	Limited by semiconductors 1200 V
Input voltage min	30–200 VDC	Below 200 VDC increase cost
Input current	Max 80 A	Above 50 A increase cost of resonance capacitors and limit by T3, T4
Range of Vin	Three times	Above two preferable variable frequency control
Load range	0–100%	Soft-switch full range
Output current	Around up 1000 A	Limited by cost of implementation secondary side of power stage, include transformer, above 500 A started dramatically increase cost
Response time	Fast	
Direction	Step-up or step-down	Bi-directional

6.6.5.1 Operation and Application Area of Resonance with PWM Topology

Now that we have clarified the disadvantages and advantages, let us consider the operational and application areas where we can use this topology. The main common application areas for both directions are high current on the low voltage side and situations where a fast response time is required, along with relatively easy achievement of low-level EMI.

Table 6.37: Possibilities for operation and application areas of the basic Resonance topology with PWM, where regulation is before the transformer.

Table 6.38 Possibilities for operation and application areas of the basic Resonance topology with PWM, where regulation is after the transformer.

Obviously, all these numbers do not need to be accepted exactly; they have tolerances depending on the target. This topology allows for changes in the control algorithm if necessary. The author used this topology for a pulse driver for laser diodes with a 166 A, 36 V output, achieving a minimum efficiency of 94% and a commutation frequency of 500 kHz.

Table 6.38 Possibility of operation and application areas of basic Resonance topology with PWM, regulation is before transformer

Power conversion	93–95%	
Co/XMFR	84–87%	
Stab/Coeff	0.8, 0.9, 1.17	0.957 average
Max frequency	1000 kHz	Limited by passive power components
Start-up method	Itself	Soft start with variable frequency control
Reliability	Middle high	
Input voltage max	850 VDC	Limited by semiconductors 1200 V
Input voltage min	10–30 VDC	Below 30 VDC increase cost
Input current	1000 A	Limited by cost of implementation secondary side of power stage, include transformer, above 500 A started dramatically increase cost
Range of Vin	Two times	
Load range	0–100%	Soft-switch full range
Output current	Max 80 A	Above 50 A increase cost of resonance capacitors and limit by T3, T4
Response time	Fast	
Direction	Step-up or step-down	Bi-directional

6.7 Dual Current-Fed: Its Modification and Development

6.7.1 How the New Topology Was Born

Around 10 years ago, the authors were tasked with developing a cost-effective, bi-directional DC–DC converter with the following specifications: a low-voltage side of 40–60 VDC and a current capacity of up to 600 A. The authors strongly believe and take full responsibility for the following statement: For the low-voltage side of isolated DC–DC converters with currents of 100–150 A and above, and for voltages below 200 VDC, one must consider one of three strategic directions: split current, Zero Current Switching (ZCS), or a combination of both.

By "split current," we mean that the current from the source is divided within the topology, rather than using two or more power stages connected in parallel. Resonant topology with PWM control is suitable for this application, but at that time, it was not cost-effective due to the challenges associated with implementing block capacitors on the low-voltage side, the power transformer, and their connection to the Printed Circuit Board (PSB). Although the situation has evolved since then, these factors still require careful consideration. In this case, the idea of splitting current is a viable approach. The dual current-fed topology and its variations are particularly attractive.

6.7.1.1 The Dual Current-Fed Topology with an Active Clamp

The dual current-fed configuration with an active clamp is a good example of this concept, as illustrated in [62, 63]. Figure 6.37 shows this topology. The main advantage of this topology is that the current through winding *W1* of the power transformer is approximately half the input current, making it significantly lower than the current through the transformer in LLC and resonant topologies with PWM. This feature has a considerable impact on the performance of the power transformer and the implementation of the entire low-voltage side of the power stage.

However, this topology also has a significant disadvantage that limits its application, which we will discuss below. Despite this, the topology can be relatively easily modified for bidirectional operation [64, 65]. This approach has been widely covered in many publications [66–68] and has inspired numerous interesting modifications [69–72]. We will examine the dual current-fed topology with active clamps. The high-voltage side of the converter is implemented with a resonant circuit and clamp diodes.

Figure 6.38 illustrates this topology. The control is managed by PWM signals for T1 and T3, which are phase-shifted by 180°. A key characteristic of this topology is that it's in three distinct modes, depending on the PWM duty cycle: below 50%, at 50%, and above 50%. We will focus on the topology's behavior at PWM duty cycles of 35, 50, and 65%.

6.7.2 *Operation of the Dual Current-Fed Topology with Active Clamp*

Let's consider the operation of the dual current-fed topology in step-up mode under the following conditions:

- **Input voltage (V1)**: 19–37 V and a maximum input current of 400 A
- **Output voltage (Vload)**: of 650 V

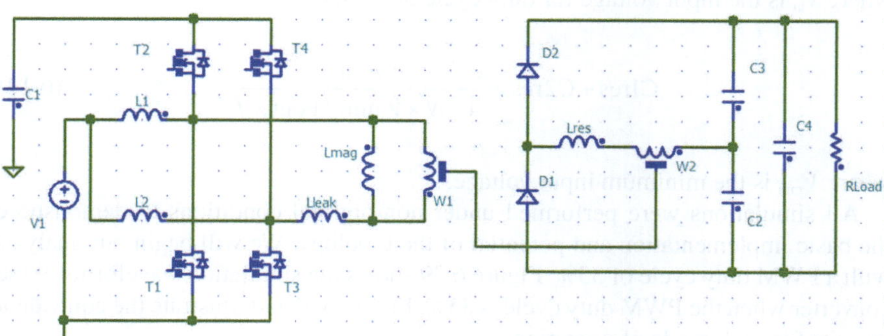

Fig. 6.37 The basic schematic of a dual current-fed topology with active clamps

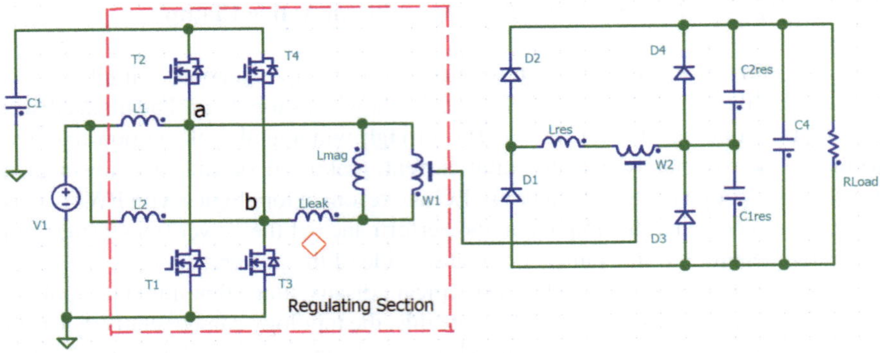

Fig. 6.38 The basic schematic of dual current fed topology with active clamps, with resonance circuit and clamp diodes

- **Maximum power**: 12 kW
- **Commutation frequency**: 100 kHz
- **Resonant inductor (Lres)**: 10 μH
- **Magnetizing inductor (Lmag)**: 3 μH
- **Leakage Inductor (Lleak)**: 100 nH
- **Resonant capacitors (C1res and C2res)**:160 nF each
- **Capacitor (C1)**: 100 μF
- **Inductors (L1 and L2)**: 4 μH each
- **The power transformer has a turns ratio**: 2:13

The preliminary calculations for the transformer's turns ratio and the resonant capacitors C1res and C2res can be determined using the following equations:

$$N = \frac{Vout}{4 \times Vin} \tag{6.11}$$

where V_{in} is the input voltage for duty cycle 50%, and

$$C1res + C2res = \frac{Pout}{4 \times N \times V\min \times Vout \times f} \tag{6.12}$$

where V_{min} is the minimum input voltage.

All simulations were performed under non-optimal conditions to demonstrate the basic implementation and potential of the topology. We will begin our analysis with a PWM duty cycle of 35%. Figure 6.39 shows the simulation waveforms of the converter when the PWM duty cycle is 35%. Figure 6.40a–d illustrate the equivalent circuits for each mode of operation.

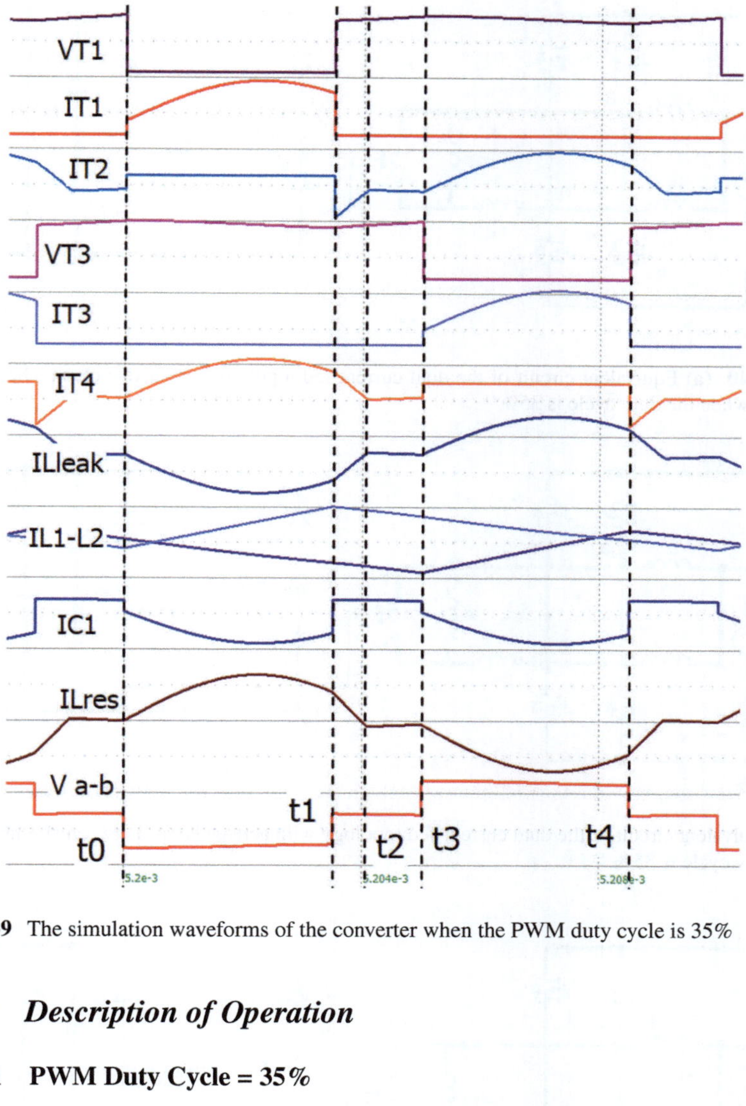

Fig. 6.39 The simulation waveforms of the converter when the PWM duty cycle is 35%

6.7.3 *Description of Operation*

6.7.3.1 PWM Duty Cycle = 35%

Starting with Fig. 6.40a, which represents the time segment t0–t1: at t0, device T1 turns on and begins conducting, while the body diode of device T2 stops conducting. The current at this moment is the combined current of inductor L1 and the magnetizing current of the power transformer. Additionally, the power conversion process begins at t0. The current from inductor L2, along with the current from capacitor C1, flows through the primary winding of the power transformer and is transferred to the secondary winding. The current in the secondary winding is directed to the load, starting from zero and rising in a sinusoidal waveform.

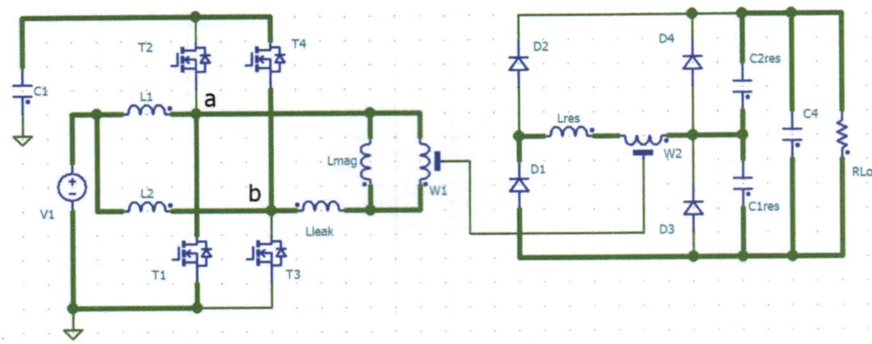

Fig. 6.40 (**a**) Equivalent circuit of the dual current-fed topology with active clamps between t0 and t1 when the duty cycle is 35%

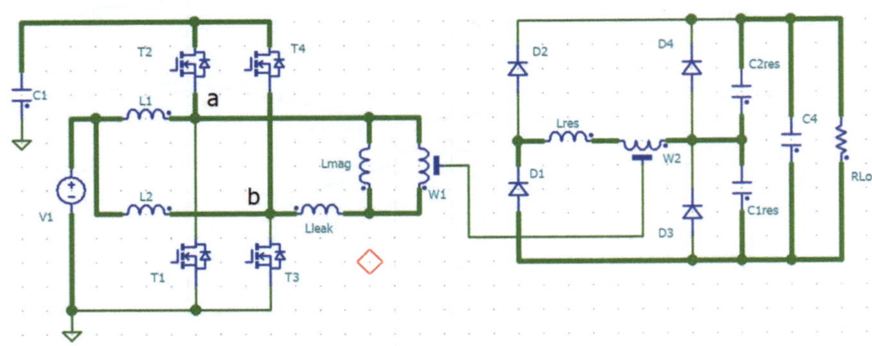

(**b**) Equivalent circuit of the dual current-fed topology with active clamps between t1 and t2 when the duty cycle is 35%

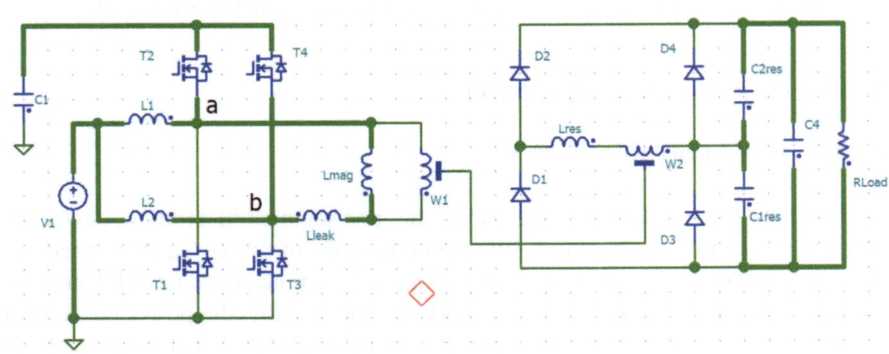

(**c**) Equivalent circuit of the dual current-fed topology with active clamps between t2 and t3 when the duty cycle is 35%

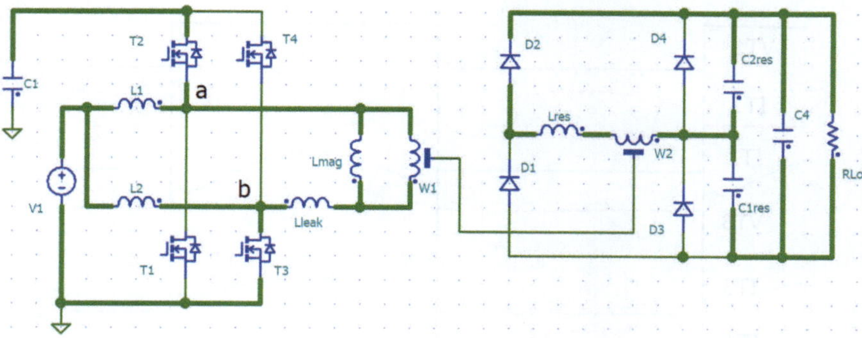

(**d**) Equivalent circuit of the dual current-fed topology with active clamps between t3 and t4 when the duty cycle is 35%

At time t1, device T1 turns off, directly interrupting the power conversion process and initiating the discharge of energy from the resonant inductors (Lleak and Lres) to the load. Figure 6.40b shows this time segment, t1–t2. By t2, the current through Lres on the high-voltage side of the power transformer reaches zero, and the power conversion process halts. As we can see, the time period between t0 and t2 encompasses the power conversion process. Additionally, at t1, capacitor C1 begins charging through the currents from inductors L1 and L2 via devices T2 and T4.

Figure 6.40c depicts the time period t2–t3. During this time, capacitor C1 continues to charge with the combined currents from L1 and L2. At t3, device T3 turns on, initiating the next half-cycle of power conversion, which mirrors the period t0–t3, with the only difference being that the current through the power transformer changes direction. Figure 6.40d shows this time segment, t3–t4.

6.7.3.2 PWM Duty Cycle = 50%

The converter's behavior at a 50% duty cycle is quite similar to its behavior at a 35% duty cycle. The primary difference is the absence of a process to discharge energy from the resonant inductors (Lleak and Lres) to the load, as there is no interruption of the resonant current.

Figure 6.41 shows simulation waveforms of the converter with a 50% PWM duty cycle, while Fig. 6.42a, b illustrate the equivalent circuits for each operation mode, highlighting the positive and negative directional currents through the power transformer (Fig. 6.42).

6.7.3.3 PWM Duty Cycle = 65%

Now, let's consider the scenario when the PWM duty cycle is 65%. Figure 6.43 shows the simulation waveforms of the converter at a 65% duty cycle, with Fig. 6.44a–d depicting the equivalent circuits for each operation mode. Starting with Fig. 6.44a, which corresponds to the time segment t0–t1:

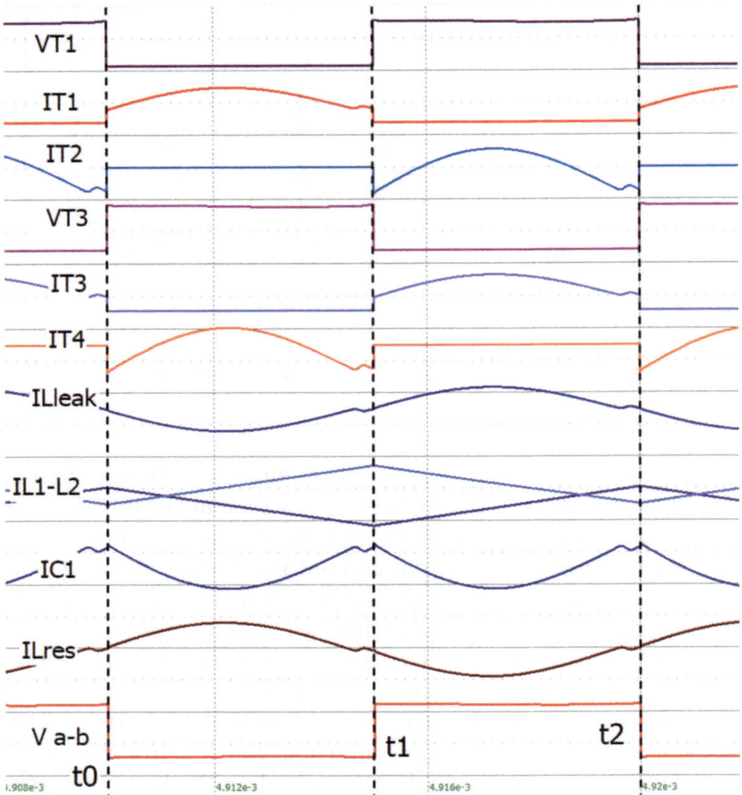

Fig. 6.41 The simulation waveforms of the converter when the PWM duty cycle is 50%

At t0, the device T1 turns off, interrupting the summed current of inductor L1 and the magnetizing current of the power transformer. This marks the beginning of the power conversion process. The current from capacitor C1 and inductor L1 starts to flow through the power transformer and device T3. This current is then transferred to the high-voltage side of the power transformer and subsequently flows to the load, starting from zero and rising in a sinusoidal waveform.

At time t1, device T1 turns on, interrupting the direct power conversion process and initiating the discharge of energy from the resonant inductors to the load. Additionally, at t1, the process of charging inductor L1 begins. Figure 6.44b illustrates this time segment t1–t2. Device T1 turns on under Zero Current Switching (ZCS) conditions or with relatively small current, as by t1, all the current from inductor L1 has been transferred to the power transformer.

At time t2, the discharge of energy from the resonant inductors to the load stops, and the process continues with current from source V1 charging inductors L1 and L2. Figure 6.44c represents this time segment t2–t3. At time t3, device T3 turns off, starting a new half-cycle of direct power conversion, which is identical to the time

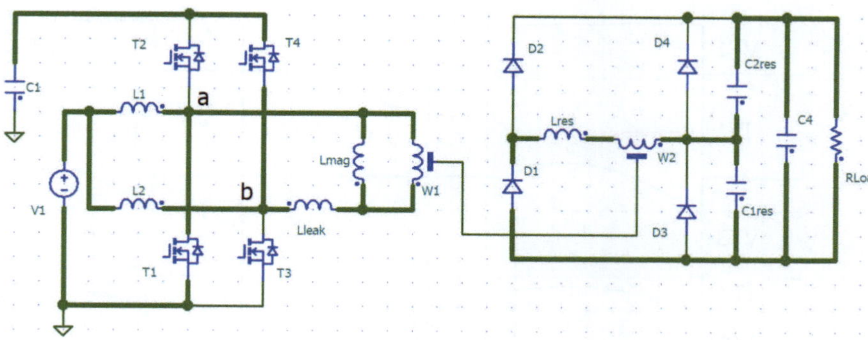

Fig. 6.42 (**a**) Equivalent circuit of the dual current-fed topology with active clamps between t0 and t1 when the duty cycle is 50%

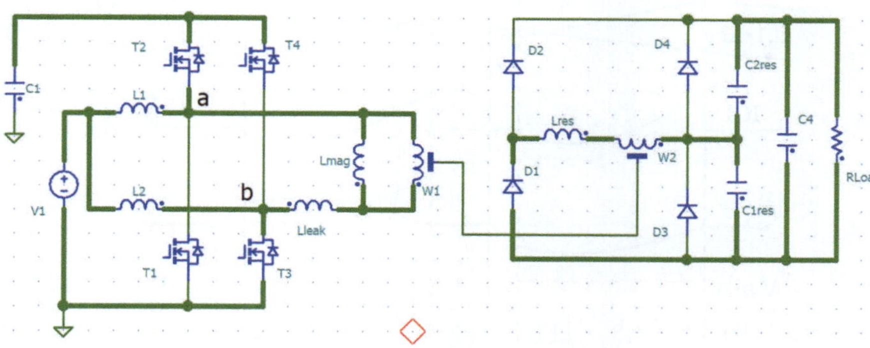

(**b**) Equivalent circuit of the dual current-fed topology with active clamps between t1 and t2 when the duty cycle is 50%

period t0–t1, except that the current direction through the power transformer changes. Figure 6.44d illustrates this time segment t3–t4.

6.7.4 Simulation and Calculation with Real Components

Now that we understand how this topology works, let's consider its performance behavior with real components, specifically semiconductors. Parasitic inductances, such as those from the PCB and bond wires of the semiconductors, are not included in this analysis. In this case, T1 and T3 each consist of four IAUT150N10S5N035 transistors connected in parallel, while T2 and T4 each consist of two IAUT150N10S5N035 transistors connected in parallel. The diodes D1–D4 are S4D30120H. The DC resistance (Rdc) of inductors L1 and L2 is 0.53 mΩ.

Table 6.39 presents the losses of the main components and the total losses of the dual current-fed converter with active clamp. Table 6.40 shows the switching stress and conduction stress on the main power components T1,T3. As can be seen from

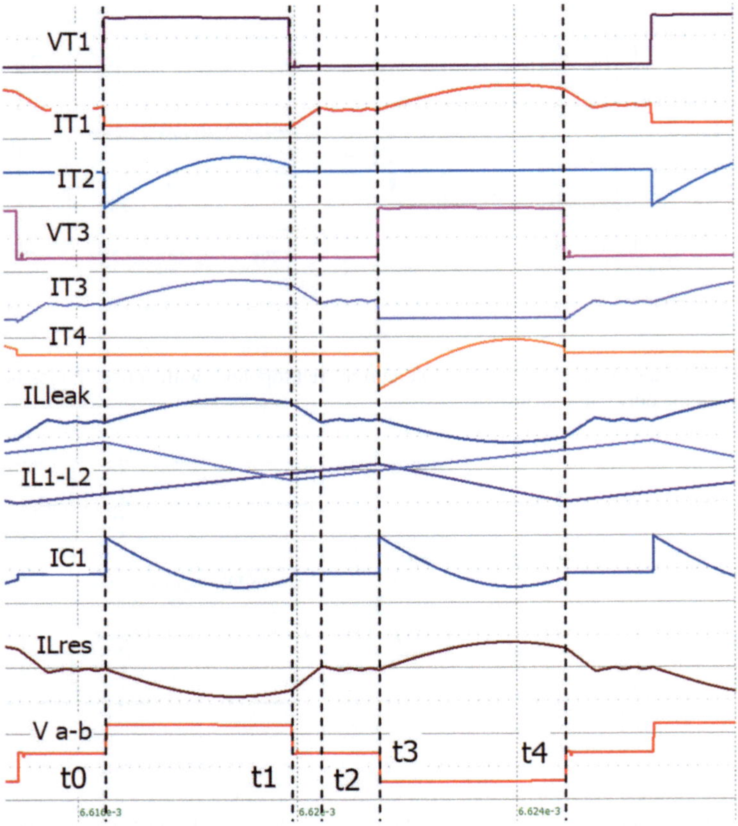

Fig. 6.43 The simulation waveforms of the converter when the PWM duty cycle is 65%

these results, on the low-voltage side, the converter belongs to the hard-switching family, while the high-voltage side operates with complete soft switching. Table 6.41 shows the delta changes in control needed to adjust Vout by 6 V when the input current is around 400 A and for different input voltages.

6.7.4.1 Specific Moments of Dual Current-Fed Topology with Active Clamp

The analysis of the data in Table 6.40 shows that the switching on the low-voltage side must consider two critical phases: turn-on and turn-off. These are crucial aspects that require deeper consideration, as they are primary factors in the switching process.

The main contributor to turn-on losses is the reverse recovery of the body diodes in devices T2 and T4, especially when the duty cycle is 50% or lower. In real-world

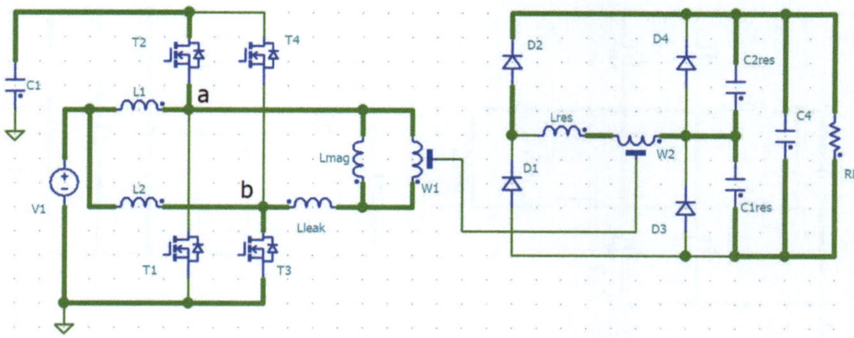

Fig. 6.44 (**a**) Equivalent circuit of the dual current-fed topology with active clamps between t0 and t1 when the duty cycle is 65%

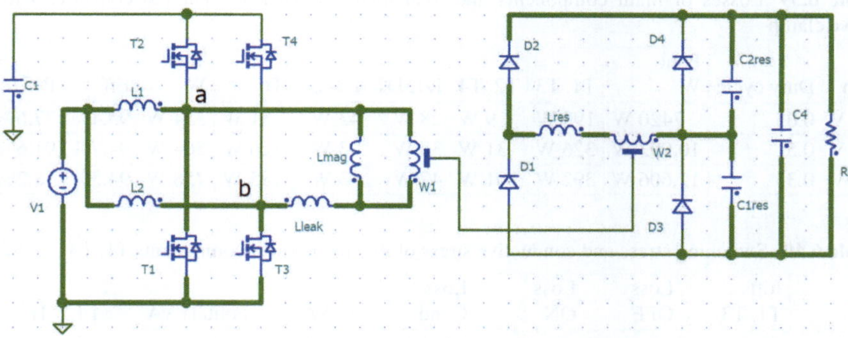

(**b**) Equivalent circuit of the dual current-fed topology with active clamps between t1 and t2 when the duty cycle is 65%

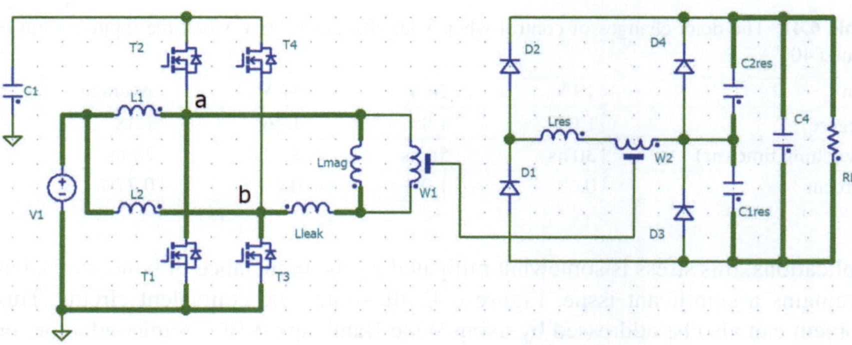

(**c**) Equivalent circuit of the dual current-fed topology with active clamps between t2 and t3 when the duty cycle is 65%

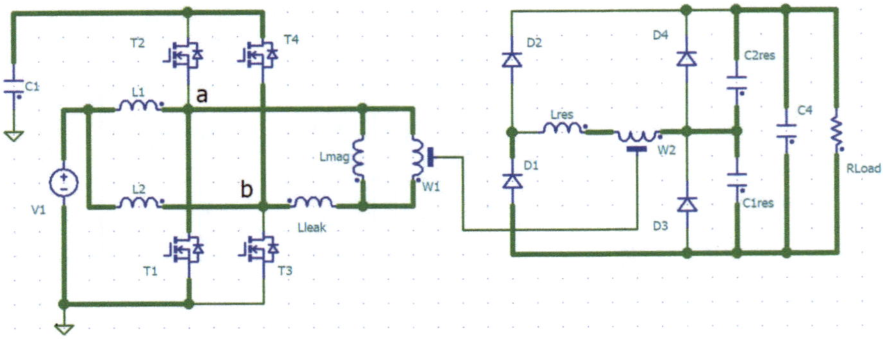

(**d**) Equivalent circuit of the dual current-fed topology with active clamps between t3 and t4 when the duty cycle is 65%

Table 6.39 Losses of main components and total losses of Dual current fed converter with active clamp

Vin	Duty cycle	Pout W	T1, T3	T2, T4	D1, D2	L1–L2	Tr	Total W	Eff	P-C
19 V	0.65	7420 W	193 W	9 W	28 W	43 W	81 W	354 W	95.4%	77.6%
26 V	0.5	10,093 W	326 W	31 W	38 W	43 W	126 W	564 W	94.7%	91.6%
37 V	0.35	12,606 W	392 W	101 W	47 W	33 W	185 W	758 W	94.3%	78.2%

Table 6.40 Switching stress and conductive stress of the main power components T1, T3

Vin	Ioff T1, T3	Loss OFF	Loss ON	Loss Cond	Tr-VA	Pout/Tr VA	I-LV Tr
19 V	224 A	32 W	18 W	143 W	8028	0.924	180 A
26 V	209 A	28 W	122 W	176 W	11,609	0.869	235 A
37 V	466 A	66 W	144 W	182 W	13,759	0.916	294 A

Table 6.41 The delta changes of control when Vout changes by 6 V when the input current is around 400 A

Vin	19 V	26 V	37 V	Average
Degree	1.08°	1.8°	1.26°	1.38°
Switching time (ns)	30 ns	50 ns	35 ns	38 ns
Percent	0.6%	1.0%	0.7%	0.77%

applications, this stress is somewhat mitigated by the inductance of bond wires, but it remains a significant issue. Figure 6.45 illustrates the equivalent circuit. This problem can also be addressed by using Wide Bandgap (WBG) semiconductors or a combination of an AC switch and a diode with very low reverse recovery energy, as shown in Fig. 6.46.

The situation with turn-off is much more complex. While the inductance of bond wires and PCB traces (as depicted in Fig. 6.45) can help during turn-on, they create

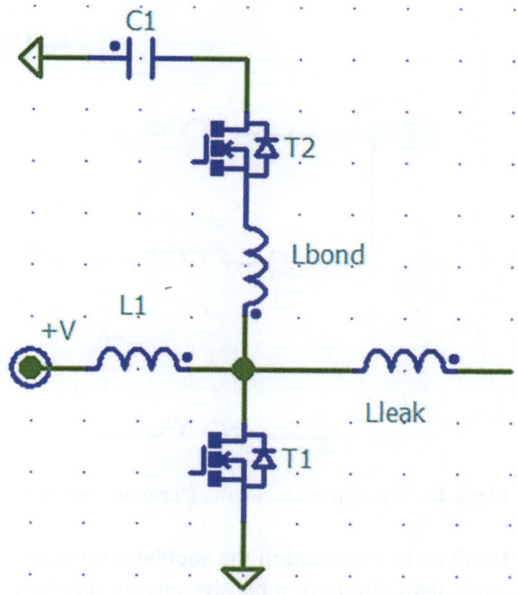

Fig. 6.45 The equivalent circuit when T1 turns-on

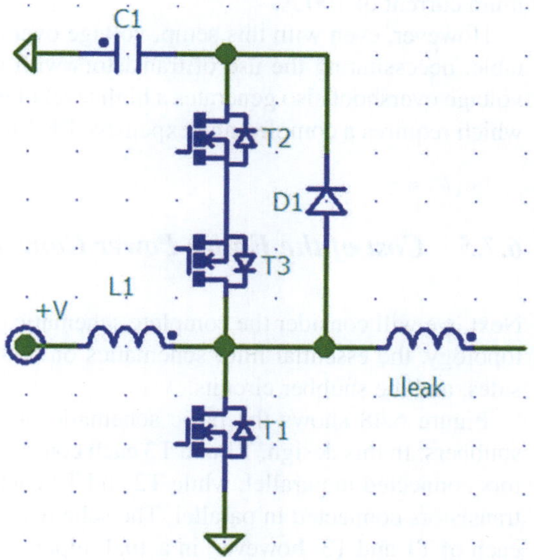

Fig. 6.46 Implementation of combination of ac switch and diode with very small revers recover energy

unfavorable conditions during turn-off. Commutation currents ranging from 200 to 400 A generate significant voltage overshoot across the transistors, necessitating the use of a snubber circuit. Passive snubbers dissipate too much energy, while energy recovery snubbers, whether passive or active, have a limited current they can manage. This limitation arises from the inductance of the bond wires in the

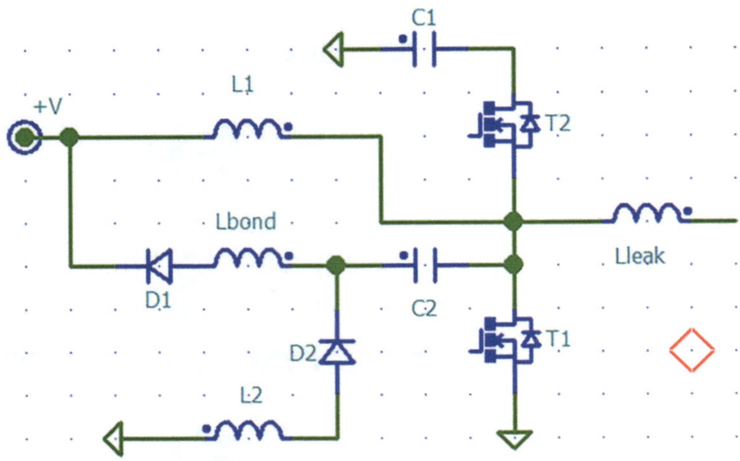

Fig. 6.47 The equivalent circuit of a passive energy recovery snubber with parasitic inductors

semiconductors used in the snubber circuit and the PCB. Figure 6.47 illustrates the equivalent circuit of a passive energy recovery snubber with parasitic inductors. In practical applications, the energy recovery snubber can efficiently handle a maximum current of 100 A.

However, even with this setup, voltage overshoot across the transistors is inevitable, necessitating the use of transistors with a much higher voltage rating. This voltage overshoot also generates a high level of electromagnetic interference (EMI), which requires a complex and expensive EMI filter.

6.7.5 *Cost of the Entire Power Converter*

Next, we will consider the complete schematic of the converter, including the basic topology, the essential filter schematics on both the high-voltage and low-voltage sides, and the snubber circuits.

Figure 6.48 shows the basic schematic of the converter, including filters and snubbers. In this design, T1 and T3 each consist of five NTBLS4D0N15MC transistors connected in parallel, while T2 and T4 each consist of three EPC2304ENGRT transistors connected in parallel. The schematic depicts a single snubber circuit for each of T1 and T3; however, in actual implementation, four snubbers are required, with each one positioned close to each NTBLS4D0N15MC transistor—in the space between these transistors.

On the low-voltage side, there is a significant level of voltage overshoot, which necessitates the use of split inductors and a common-mode filter. Table 6.42 presents the cost of the basic components for the power stage of the entire unit. Table 6.43 outlines the application areas and general performance of the dual current-fed converter with an active clamp.

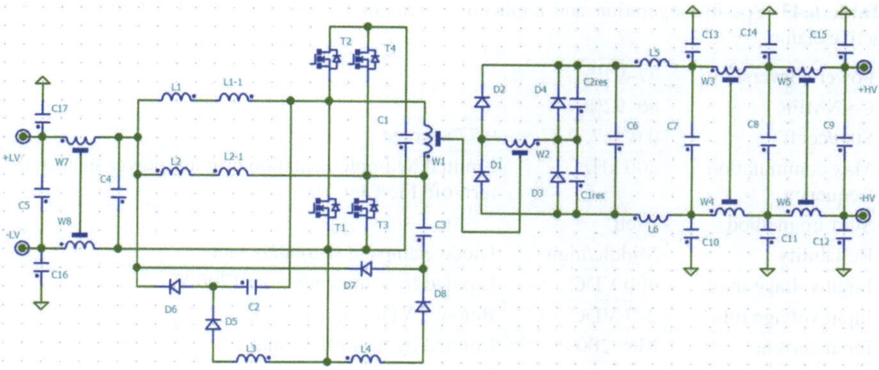

Fig. 6.48 The basic schematic of the converter, including filters and snubbers

Table 6.42 Cost of the basic components for the power stage of the whole unit

Parts	Part N	QTY	Price 1p $	Total $ Parts	Total $ Section
T1, 3	NTBLS4D0N15MC	10	3.07	30.7	
T2, 4	EPC2304ENGRT	6	4.47	26.82	
W1–W2	Power Tr without holders	1	140.0	140.0	
C1	C1812X335K1RACAUTO	80	0.42	33.6	
L1, 1-2, 2, 2-1	B82559B2102A027	16	4.31	68.96	
D1–4	S4D30120H	4	6.74	26.96	
C1res, C2res	R76UN22204040J	14	0.63	8.82	
C6	C4AQUBU4180A1XJ	3	0.99	2.97	
C4	CL32B106KBJNNNF	15	0.127	1.91	
Power supply for gate drivers T1–T4	LMT78_15-1.0R	1	2.9	2.9	
Gate driver T1–T4	UCC5390-Q1	4	1.52	6.08	
Power stage	**Total**	**1**			**349.72**
C2, 3	C1210X223KBGAC7800	8	0.58	4.64	
D5–8	MBR10U200-TP	16	0.26	4.16	
L3, 4	CSCM1256-2R7M	8	2.24	17.92	
Aux circuits	**Total**				**26.72**
C7, 8, 9	C4AQUBU4180A1XJ	3	0.99	5.94	
L5, L6	CPEX3231A-270MC	2	11.8	23.6	
W7–W8	High current common mode choke	1	60	60.0	
C5	CL32B106KBJNNNF	10	0.127	1.27	
W3, 4, 5, 6	B82726S6243A040	2	9.4	18.8	
Filter	**Total**				**109.61**
Summary	Power stage	Aux	Filters		486.05
Coefficient	On big size of parts	1.4			194.42
	Total cost of converter				**680.47**

Table 6.43 Possible operation and application areas of basic Dual current fed converter with active clamp

Power conversion	77–92%	
Co/XMFR	86–92%	
Stab/coeff	0.6, 0.7, 0.77	0.77 average
Max commutation frequency	300 kHz	Limited by implementation primary power stage and turn-off T1, T3
Start-up method	Itself	
Reliability	Middle/high	Diode clamps in secondary side
Input voltage max	400 VDC	Limited by semiconductors 1200 V
Input voltage min	2–7 VDC	Below 7 VDC increase cost
Input current	Max 200 A	Limited by turn-off T1 and T3
Range of vin	1.8 times	Limited by implementation primary side of power stage
Load range	0–100%	
Output current	Around up 500 A	Limited by cost of implementation secondary side of power stage, include transformer
Response time	Fast/middle	
Direction	Step-up or step-down	Bi-directional. Replace output diodes to active switches

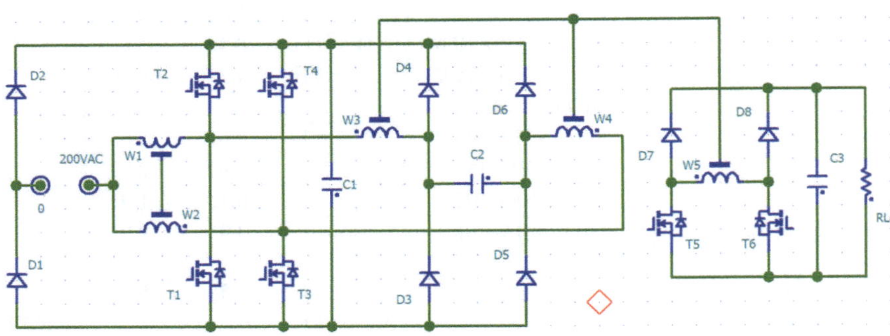

Fig. 6.49 The schematic of the power stage for this AC-DC converter

In conclusion, the dual current-fed converter with an active clamp is suitable for applications where the input current does not exceed 200 A. This explains why, in most publications, the authors typically use input currents of no more than 50 A and a commutation frequency of 50 kHz with a duty cycle above 50%.

Another application area for this topology is in isolated AC-DC converters. As illustrated in references [73, 74], the authors employed this topology in an isolated AC-DC converter achieving a power factor of 0.99 and delivering 20 kW of power. Figure 6.49 presents the schematic of the power stage for this AC-DC converter.

Upon analyzing the results, it becomes clear that the basic dual current-fed topology with an active clamp is not suitable for applications requiring 400 A input current. However, there is a strong temptation to continue working with this topology.

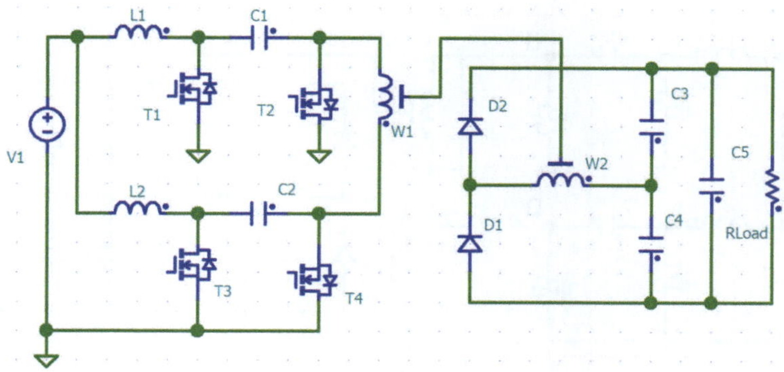

Fig. 6.50 The basic schematic for implementing innovated idea

6.7.5.1 New Approach to Enhance Dual Current-Fed Topology

The authors propose an innovative method to significantly reduce or even eliminate the limitations of traditional designs by enabling the low-voltage side to handle input currents well beyond 200 A. The principle behind this idea is that transistors T1 and T3, as shown in Fig. 6.50, must follow a specific rule during turn-off: the higher the commutation current, the lower the commutation voltage must be. When the current reaches its maximum, the commutation should ideally achieve ZVS or be very close to it. In essence, the authors aimed to combine the advantages of two topologies—resonant with PWM and dual current-fed with active clamp—into a single, unified topology. Figure 6.50 shows the basic schematic for implementing this idea.

The primary difference between the dual current-fed topology with an active clamp and the new topology is in how energy is managed by the input inductors L1 and L2. In the dual current-fed topology, L1 and L2 receive energy directly from the source and then discharge it to the power transformer and capacitor C1. During this discharge process, L1 and L2 are connected in series with the energy source, allowing energy to move directly from the source to the load.

In contrast, in the new topology, inductors L1 and L2 primarily charge capacitors C1 and C2. After charging, C1 and C2 discharge into the transformer and provide Zero Voltage Switching (ZVS) turn-off for switches T1 and T3. In this design, energy is exchanged between inductors L1, L2 and capacitors C1, C2, without any direct power conversion from the source to the load. Therefore, this topology is referred to as dual current-fed with switched capacitors.

Figure 6.51 illustrates the basic schematic of the dual current-fed with switched capacitors topology, incorporating resonance and clamp diodes, as described in [75]. This topology shares several characteristics with the dual current-fed with active clamp design. Therefore, we will analyze the dual current-fed with switched capacitors topology using the same methodology.

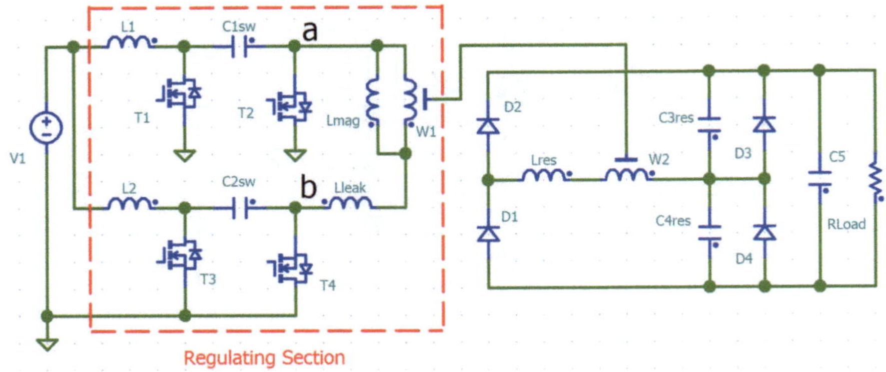

Fig. 6.51 The basic schematic of the dual current-fed with switched-capacitors topology

We will consider the behavior of the topology under PWM duty cycles of 35, 50, and 65%.

6.7.5.2 Operation of the Dual Current-Fed Topology with Switch Capacitors

Let's consider the operation of the dual current-fed topology with switch capacitors under the following conditions step-up mode:

- **Input voltage (V1)**: 19–37 V and a maximum input current of 400 A
- **Output voltage (Vload)**: of 650 V
- **Maximum power**: 12 kW
- **Commutation frequency**: 100 kHz
- **Resonant inductor (Lres)**: 13 μH
- **Magnetizing inductor (Lmag)**: 10 μH
- **Leakage inductor (Lleak)**: 100 nH
- **Switch capacitors (C1sw and C2sw)**: 13.3 μF each
- **Resonant capacitors (C1res and C2res)**:160 nF each
- **Capacitor (C1)**: 100 μF
- **Inductors (L1 and L2)**: 4 μH each
- **The power transformer has a turns ratio**: 2:14

All simulations were conducted under non-optimal conditions to demonstrate the basic functionality and potential of the topology. We begin with a PWM duty cycle of 35%. Figure 6.52 shows the simulation waveforms of the converter at this duty cycle. Figure 6.53a–d illustrate the equivalent circuits for each operation mode.

Starting with Fig. 6.53a, which depicts the time segment t0–t1: at t0, device T1 turns on and begins conduction, causing the body diode of device T2 to stop conducting. At this moment, capacitor C1sw ceases charging and begins discharging into the primary winding of the power transformer. The energy stored in capacitor C1sw is then transferred to the load and used to charge the resonant inductors (Lleak

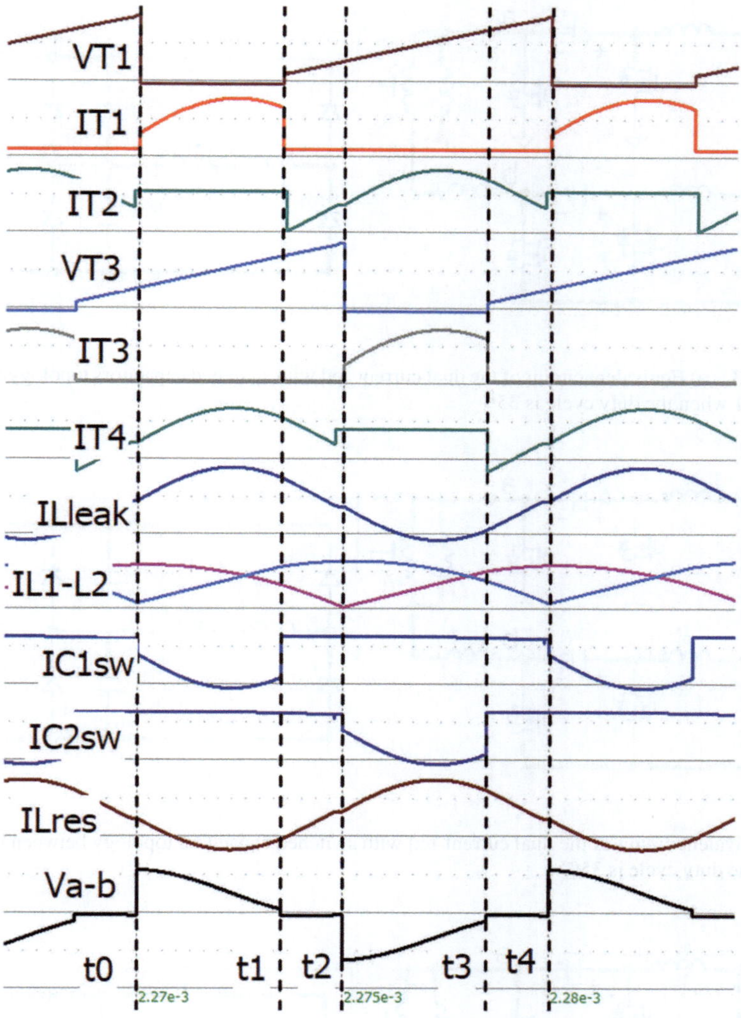

Fig. 6.52 The simulation waveforms of the converter at 35% duty cycle

and Lres). This marks the beginning of direct power conversion from capacitor C1sw during this half cycle.

At time t1, device T1 turns off, interrupting the direct power conversion process and initiating the discharge of energy from the resonant inductors (Lleak and Lres) to the load. When T1 turns off, it interrupts the current, which is the sum of the current from L1 and the current flowing through capacitor C1sw. The voltage across T1 at this moment is determined by the voltage across capacitor C1sw at the time of interruption. This voltage is influenced by the capacitance value and the current discharging the capacitor—higher current results in lower voltage and reduced

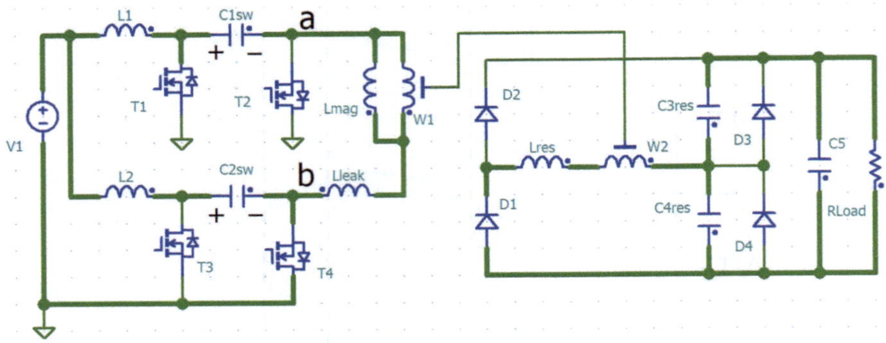

Fig. 6.53 (**a**) Equivalent circuit of the dual current-fed with switched capacitors topology between t0 and t1 when the duty cycle is 35%

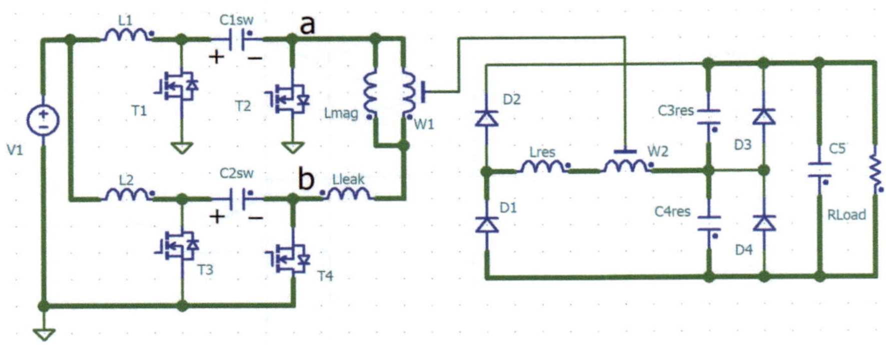

(**b**) Equivalent circuit of the dual current-fed with switched capacitors topology between t1 and t2 when the duty cycle is 35%

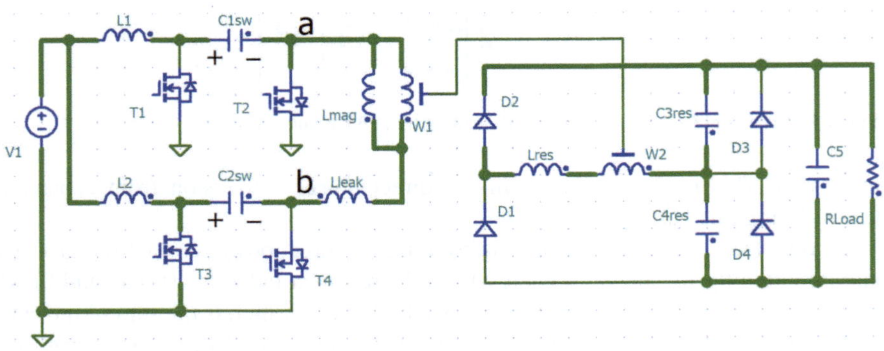

(**c**) Equivalent circuit of the dual current-fed with switched capacitors topology between t2 and t3 when the duty cycle is 35%

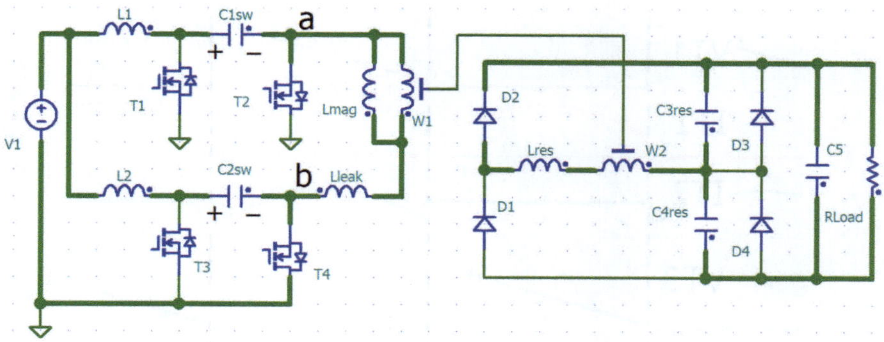

(**d**) Equivalent circuit of the dual current-fed with switched capacitors topology between t3 and t4 when the duty cycle is 35%

switching losses. If the capacitor fully discharges, T1 and T3 could achieve pure Zero Voltage Switching (ZVS). Figure 6.13b illustrates this time segment t1–t2.

Also, at time t1, capacitor C1sw stops discharging and begins to recharge via inductor L1. During the period t0–t2, capacitor C2sw is charged by inductor L2.

At time t2, device T3 turns on, initiating the next half cycle of power conversion, where capacitor C2sw begins discharging into the primary winding of the power transformer and charging the resonant inductors (Lleak and Lres). Figure 6.53c shows this time segment t2–t3.

At time t3, device T3 turns off, stopping the discharge process of capacitor C2sw to the load and starting the discharge process of the resonant inductors to the load. Figure 6.53d illustrates this time segment t3–t4. Figure 6.54 shows the simulation waveforms of the converter operating with a 50% PWM duty cycle, while Fig. 6.55a, b illustrate the equivalent circuits for each operation mode, depicting both positive and negative directional current through the power transformer. As we can see, these waveforms are very similar to those observed in the dual current-fed topology with an active clamp, with one key difference: the voltage during the turn-off of devices T1 and T3. In the switched capacitor topology, turn-off switching losses are significantly reduced, resulting in a much lower level of switching losses.

switched capacitor topology, turn-off switching losses are significantly reduced, resulting in a much lower level of switching losses.

Next, we consider the converter operation with a PWM duty cycle of 65%. Figure 6.56 shows the simulation waveforms for this duty cycle, while Fig. 6.57a–d depict the equivalent circuits for each operation mode. Starting with Fig. 6.57a, which represents the time segment t0–t1: at t0, device T1 turns on and begins conducting, causing the body diode of device T2 to stop conducting. At this moment, capacitor C1sw stops charging and begins discharging into the primary winding of the power transformer. The energy from capacitor C1sw is transferred to the load and charges the resonant inductors (Lleak and Lres), initiating the direct power conversion process from capacitor C1sw in this half cycle. During this time segment (t0–t1), capacitor C2sw continues discharging. The current through the resonant inductors is determined by the voltage difference between capacitors C1sw and C2sw.

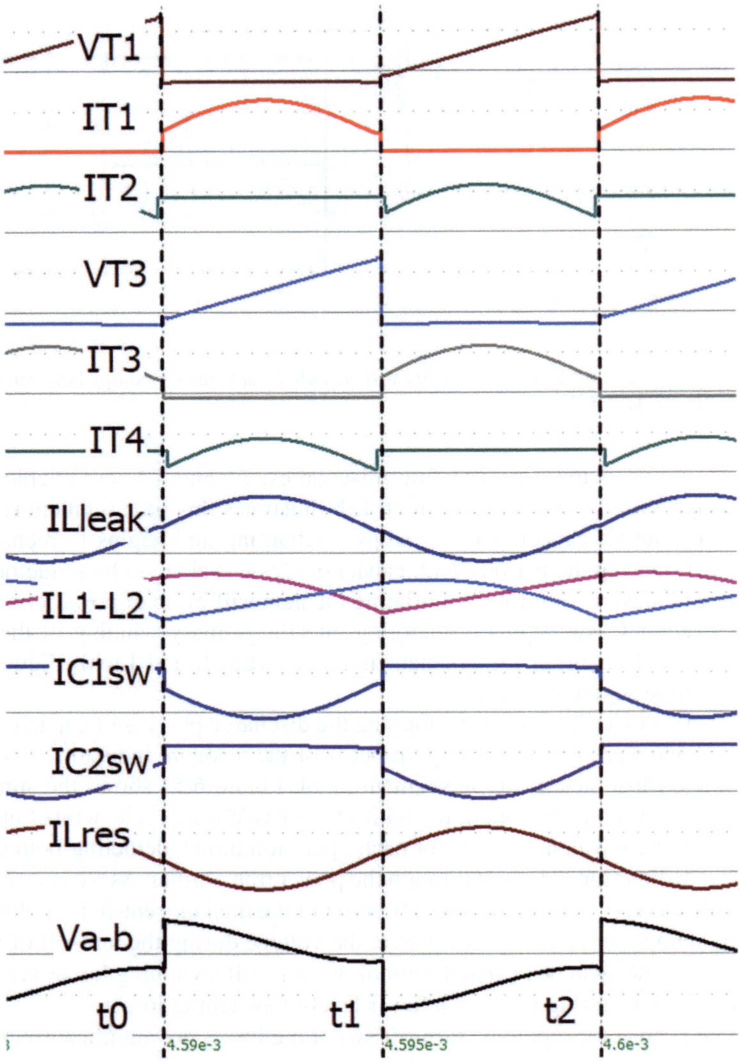

Fig. 6.54 The simulation waveforms of the converter at 50% duty cycle

At time t1, device T3 turns off, causing capacitor C2sw to stop discharging and start recharging through the current from inductor L2. During the period t1–t2, only capacitor C1sw discharges into the power transformer, as shown in Fig. 6.57b.

At time t2, device T3 turns on again, and capacitor C2sw stops charging and begins discharging into the power transformer, starting a new half cycle of power conversion. This cycle is analogous to the period t0–t2, except that the direction of the current through the resonant inductors changes. Figure 6.57c illustrates this time segment t2–t3.

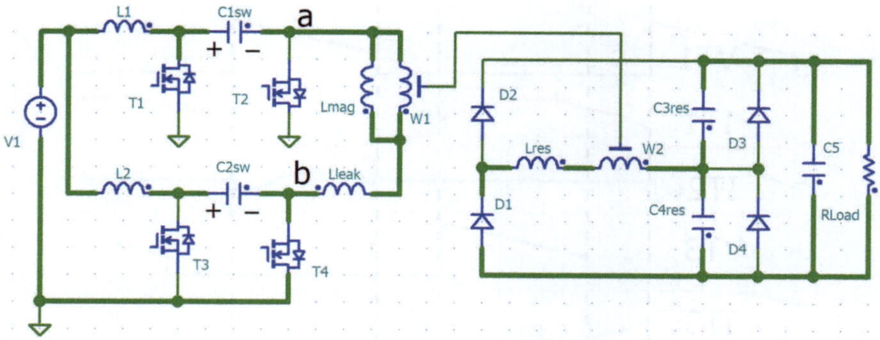

Fig. 6.55 (**a**) Equivalent circuit of the dual current-fed with switched capacitors topology between t0 and t1 when the duty cycle is 50%

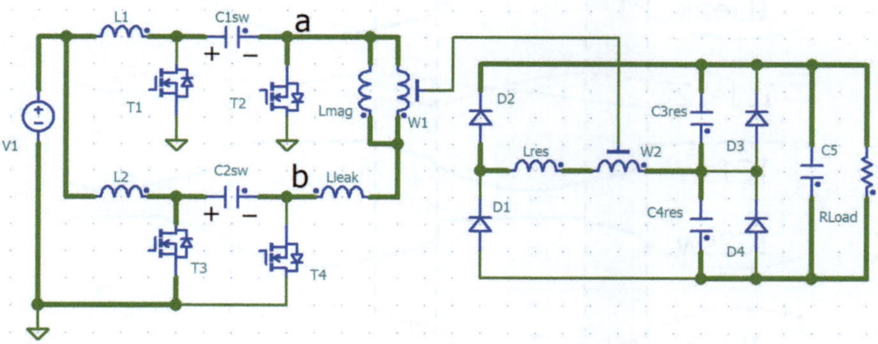

(**b**) Equivalent circuit of the dual current-fed with switched capacitors topology between t1 and t2 when the duty cycle is 50%

At time t3, device T1 turns off, causing capacitor C1sw to stop discharging and begin recharging through the current from inductor L1. At time t4, device T1 turns on, ending this half cycle of power conversion and starting the next half cycle. Figure 6.57d shows this time segment t3–t4.

6.7.5.3 Calculation of Main Power Components

Now that we understand how this topology operates, let's consider how to choose the power transformer ratio, power capacitors, and resonant inductor. The voltage gain of this converter, under ideal conditions and assuming an ideal transformer, can be calculated using the following equation for a duty cycle $D = 0.5$:

$$\frac{Vout}{Vin} = \frac{2N}{1-D} \tag{6.13}$$

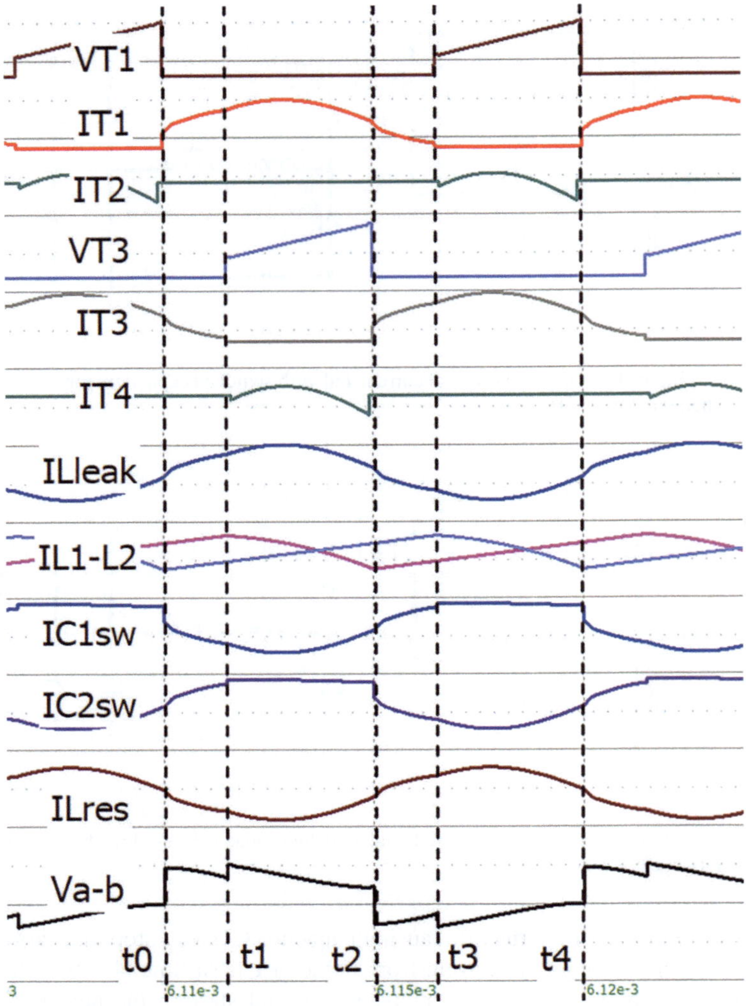

Fig. 6.56 The simulation waveforms of the converter at 65% duty cycle

where N is the transformer turns ratio $N = W2/W1$.

For practical engineering calculations when D ranges between 0.3 and 0.7, the voltage gain can be approximated using the following equation:

$$\frac{Vout}{Vin} = \frac{2N\sqrt{\frac{Ton}{T}}}{1-D} \tag{6.14}$$

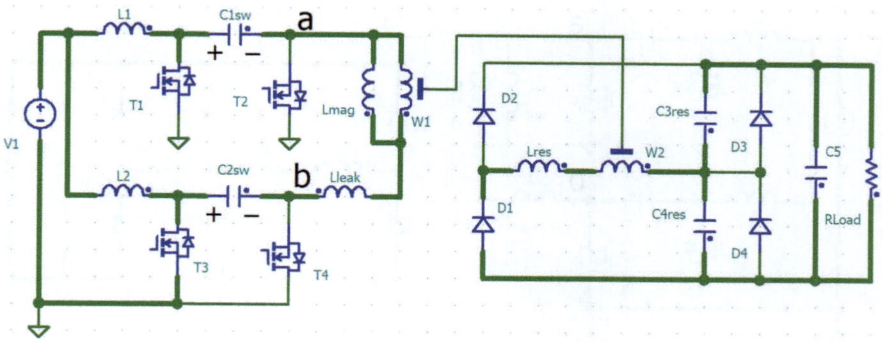

Fig. 6.57 (**a**) Equivalent circuit of the dual current-fed with switched capacitors topology between t0 and t1 when the duty cycle is 65%

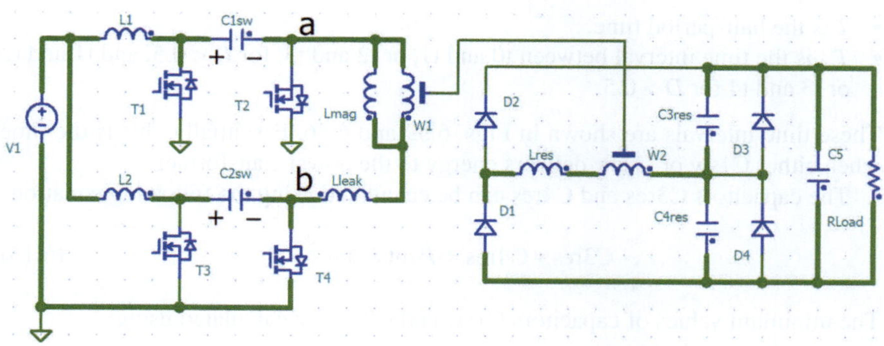

(**b**) Equivalent circuit of the dual current-fed with switched capacitors topology between t1 and t2 when the duty cycle is 65%

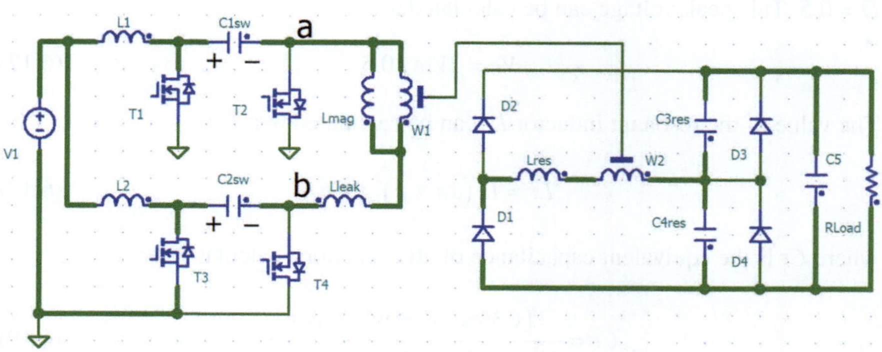

(**c**) Equivalent circuit of the dual current-fed with switched capacitors topology between t2 and t3 when the duty cycle is 65%.

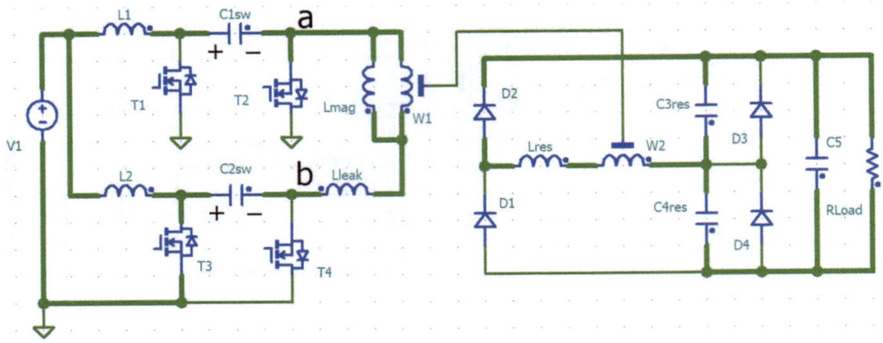

(**d**) Equivalent circuit of the dual current-fed with switched capacitors topology between t3 and t4 when the duty cycle is 65%

- T is the half-period time.
- T_{on} is the time interval between t0 and t1, or t2 and t3, for $D < 0.5$, and t1 and t2 or t3 and t4 for $D > 0.5$.

These time intervals are shown in Figs. 6.52 and 6.56. Essentially, this is the time when either C1sw or C2sw delivers energy to the power transformer.

The capacitors C3res and C4res can be calculated using the following equation:

$$\text{C3res} + \text{C4res} = Pout / Vout^2 \times f \tag{6.15}$$

The minimum values of capacitors C1sw and C2sw are calculated using:

$$\text{C1sw} + \text{C2sw} = Pout / Vc^2 \times f \tag{6.16}$$

where Vc is the peak voltage across the capacitors at maximum output power with $D = 0.5$. This peak voltage can be calculated as:

$$Vc = 2Vin / 0.5 \tag{6.17}$$

The value of the resonant inductor Lr can be calculated by:

$$Lr = 1 / (2\pi \times f)^2 \times Cr \tag{6.18}$$

where Cr is the equivalent capacitance of all capacitors, calculated as:

$$Cr = \frac{(\text{C3res} + \text{C4res}) \times \text{C1sw}}{N^2 \times (\text{C3res} + \text{C4res}) + \text{C1sw}} \tag{6.19}$$

Note: C1sw = C2sw.

6.7.6 Simulation and Calculation with Real Components

We now consider the results of simulations conducted under various conditions, including different input voltages and load levels. Figure illustrates the schematic used for both simulation and calculation. The component values are as follows: L1 and L2 are 4 μH, Rdc is 0.53 mΩ, Lleak is 0.1 μH, Lmag is 10 μH, C1sw and C2sw are 13.2 μF, Lres is 12 μH, and C3res and C4res are 150 nF. The switches T1 and T3 each consist of five EPC2304 devices connected in parallel, while T2 and T4 each consist of three EPC2304 devices connected in parallel. The diodes D1 through D4 are S4D30120H. The power transformer has a turns ratio of 7, and the commutation frequency is 100 kHz. The transformer's leakage inductance is negligible, so it does not significantly influence the converter's behavior and is therefore not included in the parts list. All simulations were performed under non-optimal conditions to demonstrate the basic implementation and potential of the topology.

Table 6.44 presents the losses of the main components and the total losses when the input current is approximately 400 A. Table 6.45 details the switching and conduction stresses on T1, T3, and other key power components. In a real-world application, it's important to account for conduction losses in the PCB, busbars, and filters. In this topology, the losses in the busbars and PCB are relatively smaller compared to resonance with PWM and LLC topologies because the current is split on the low-voltage side.

As observed, the primary power losses occur in the transistors and the power transformer. However, there is potential for improvement if needed. The current on the low-voltage side of the power transformer is around 300 A, which is less than 500 A, indicating that there is room for enhancement. Conduction losses are the main source of losses in the transistors, while switching losses are significantly

Table 6.44 Losses of main components and total losses of Dual current fed converter with switch capacitors

Vin	Duty cycle	Pout W	T1, T3	T2, T4	D1, D2	L1–L2	Tr	Total W	Eff	P-C
19.5 V	0.65	7670 W	144 W	11 W	38 W	45 W	82 W	320 W	96%	96%
24 V	0.5	9770 W	194 W	31 W	46 W	47 W	115 W	433 W	95.7%	94%
30 V	0.41	12,109 W	226 W	72 W	56 W	47 W	160 W	561 W	95.6%	94%
37 V	0.35	11,820 W	188 W	105 W	55 W	30 W	159 W	537 W	95.6%	93%

Table 6.45 Switching, conductive stress of T1, T3, and main power components

Vin	Ioff T1, T3	Loss OFF	Loss ON	Loss Cond	Tr-VA	Pout/Tr VA	I-LV Tr	Irms C1sw, C2sw
19.5 V	21 A	Zero	2 W	142 W	9040	0.848	197 A	194 A
24 V	187 A	5 W	17 W	172 W	12,490	0.81	243 A	230 A
30 V	380 A	6 W	20 W	200 W	16,225	0.746	295 A	265 A
37 V	472 A	23 W	12 W	153 W	16,060	0.736	294 A	242 A

Table 6.46 Different power level with varying input voltage

Vin	Pout W	Eff	Pout W	Eff
19 V	3807 W	97.5%	1096 W	97.1%
22 V	5013 W	97.4%	1192 W	97.3%
29 V	6110 W	97.3%	1434 W	97.5%
36 V	7135 W	97.0%	1864 W	97.4%

Table 6.47 Results under overload conditions

Vin	Duty cycle	Pout W	Total W	Eff	I-LV Tr	Irms C1sw, C2sw
19.5 V	0.65	9593	552 W	94.5%	260 A	258 A
24 V	0.5	11,008	609 W	94.7%	293 A	270 A
30 V	0.41	11,892	617 W	95.0%	309 A	267 A
37 V	0.35	11,038	556 W	95.2%	300 A	233 A

Table 6.48 The delta changes of control for change Vout on 6 V when the input current around 400 A

Vin	19.5 V	24 V	30 V	37 V	Average
Degree	4.68°	2.52°	1.08°	1.08°	2.34°
Switching time (ns)	130 ns	70 ns	30 ns	30 ns	65 ns
Control percentage	2.6%	1.4%	0.6%	0.6%	1.3%

lower. To improve this, we can increase the number of parallel transistors and add a simple energy recovery snubber circuit.

Table 6.46 shows the performance at different power levels with varying input voltages, while Table 6.47 presents the results under overload conditions, where the load resistance is reduced to half of the maximum load. As seen, due to the converter's controlled energy transfer characteristic, the output power does not increase under overload conditions. Table 6.48 shows the delta changes in control required to adjust Vout by 6 V when the input current is around 400 A and the input voltage varies.

The authors implemented this topology with an input current of up to 600 ADC on the low-voltage side, achieving an overall efficiency of 94–95% at full power. The reported losses encompass the PCB, power connections, busbars, filters, and other components. Now, we will analyze the entire converter schematic, including the basic topology and the filter schematics on both the high-voltage and low-voltage sides. Figure 6.58 illustrates the basic schematic of the converter with filters.

On the low-voltage side, an autotransformer is employed. The primary reason for this choice is the minimal or negligible overshooting and low EMI on the low-voltage side, eliminating the need for split inductors as required in the previous topology shown in Fig. 6.48. However, the inductors in this application exhibit significant losses, including DC losses, AC losses, and core losses, with DC losses

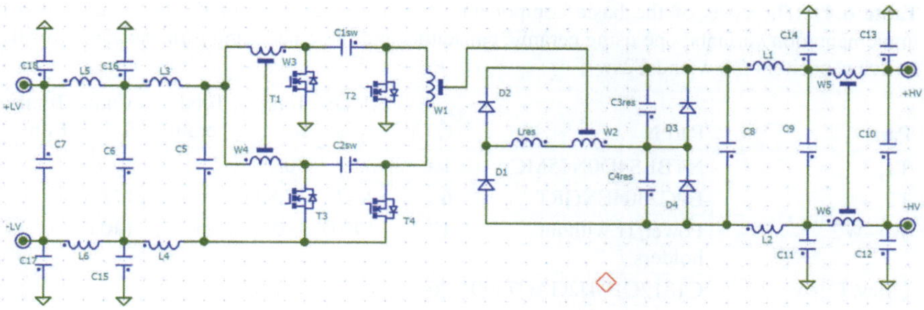

Fig. 6.58 The basic schematic of the converter with filters

accounting for approximately 30–50% of the total losses. By using an autotransformer, we achieve lower total losses, reduced size, and lower costs for mass production.

6.7.6.1 Cost of the Entire Power Converter

On the high-voltage side, a single stage of a common-mode filter is sufficient, as there is no interruption in the current, and soft-switching with ZCS is employed. The dV/dt is around 2500–3000 V/µs, with a power conversion efficiency of approximately 94%. Table 6.49 compares the costs of the basic components for the power stage of the entire unit in two implementation variants: one using ceramic capacitors and the other using film capacitors for the switch capacitors C1sw and C2sw. This topology is clearly workable and suitable for the application, with an input current of 400 A. However, it does come with both advantages and disadvantages.

6.7.7 Disadvantages and Advantages

The primary disadvantage of this topology lies in the capacitors, specifically the switching capacitors (Csw). The current flowing through these capacitors is exceedingly high, which limits the applicability of this topology due to cost considerations. Ceramic capacitors, particularly NPO or COG types, offer the best technical performance for this application, but they come at a high cost. For instance, the ceramic capacitor C1812C104J2GAC7800 costs $147.84 (264p x $0.56 = $147.84). In comparison, the film capacitor R76PN34705050J (56p) costs $47. This presents a trade-off: while film capacitors are less expensive, they are larger in size. On the other hand, ceramic capacitors, although more costly, are much smaller and allow for a higher switching frequency, up to 400 kHz or more. This disadvantage is typical for any topology that relies on switching capacitors, necessitating careful selection by the design engineer.

Table 6.49 The costs of the basic components for the power stage of the entire unit in two implementation variants: one using ceramic capacitors and the other using film capacitors for the switch capacitors C1sw and C2sw

Parts	Part N	QTY	Price 1p $	Total $	Total ceram	Total $	Total Film
T1, 3	NTBLS4D0N15MC	10	3.07	30.7		30.7	
T2, 4	EPC2304ENGRT	6	4.47	17.88		17.88	
W1–W2	Power Tr without holders	1	140.0	140.0		140.0	
C1sw, C2sw	C1812C104J2GAC7800	264	0.56	147.84			
	R76PN34705050J	56	0.84			47.0	
W3–W4	Autotransformer	1	40	40		40	
D1–4	S4D30120H	4	6.74	26.96		26.96	
Lres	E58/11/38	4	3.3	13.2		13.2	
C3res, C4res	R76UN22204040J	14	0.63	8.82		8.82	
C5	CL32B106KBJNNNF	20	0.127	2.54		2.54	
C8	C4AQUBU4180A1XJ	3	0.99	2.97		2.97	
Power supply for gate drivers T1–T4	LMT78_15-1.0R	1	2.9	2.9		2.9	
Gate driver T1–T4	UCC5390-Q1	4	1.52	6.08		6.08	
Power stage	**Total**				**439.89**		**339.05**
C9, 10	C4AQUBU4180A1XJ	3	0.99	5.94		5.94	
L1, L2	CPEX3231A-270MC	2	11.8	23.6		23.6	
W5, W6	B82726S6243A040	1	9.4	9.4		9.4	
L3–6	00K4119U090	8	1.5	12		12	
C6–7	CL32B106KBJNNNF	15	0.127	3.81		3.81	
Filters	**Total**				**54.75**		**54.75**
Summary		Power	Stage	Filters	494.64		393.8
Coefficient	On big size of parts	1.3			148.39		
Coefficient	On big size of parts	1.35					137.83
	Total cost of converter				**643.03**		**531.63**

Another drawback is the relatively high voltage across transistors T1–T4, which is approximately four times higher than the average input voltage. Additionally, the recovery time of the body diodes in transistors T2 and T4 poses a challenge. Although this topology operates close to Zero Voltage Switching (ZVS) mode, transistors T1 and T3 still interrupt high currents, requiring careful attention to the layout of the PCB in the power stage.

The last point of concern is the relatively large resonant inductor. This component has a dual effect: while it improves the behavior of the power stage by increasing the duty cycle of power conversion (P-C), it also complicates implementation due to the need for an additional inductor. Moreover, this inductor impacts the control response time, leading to an increase.

The primary advantage of this topology is that the current through the low-voltage side winding of the power transformer is approximately half as much, while the voltage across this winding is roughly twice as high compared to, for example, a resonance with PWM configuration. This performance characteristic is crucial because the current through the transformer winding is AC, which facilitates cost reduction in the transformer itself, simplifies the mounting of the power transformer, and improves efficiency by approximately 0.25–0.75%. On the low-voltage input side, this configuration allows for a higher number of turns in the winding—potentially two turns instead of one. This topology belongs to the family of dose energy transfer converters, and it specifically offers the possibility of double dosing. The first dosing occurs via resonance capacitors with clamp diodes on the high-voltage side, and the second is through the switching capacitors, provided they are optimized for this purpose (i.e., they fully discharge). This capability enhances the converter's reliability and makes it less sensitive to short circuits on the load. Moreover, it allows these converters to be connected in parallel without issues related to current sharing; the current will naturally balance itself based on the converter's parameters without requiring control compensation. Another advantage is the relatively high power conversion efficiency (P-C) across the full power range with a soft-switch rectifier. This helps reduce EMI levels and simplifies the filter design, leading to further cost reductions.

6.8 Conclusions

Having discussed the advantages and disadvantages of this topology, we can now consider its potential application areas. The primary application for this topology is in scenarios where a high input current at low voltage is required, along with a high-voltage output. It is particularly suited for situations where a relatively low EMI level, moderate response time, and cost efficiency are important. Table 6.50 outlines the application areas and general performance characteristics of the dual current-fed converter with switching capacitors. It should be noted that the values provided in the table are approximate and may vary depending on the specific objectives of the application.

This topology also offers the flexibility to adjust the control algorithm if needed. Further details about this topology have been published in references [76–80], and it is covered by patents [81–84]. Additionally, this topology has been successfully implemented as the core of a three-port bidirectional isolated DC–DC converter for the electrification of transportation units.

Finally, the author wishes to emphasize that this chapter illustrates how an engineering vision evolves and culminates in a new technical solution.

Table 6.50 Possible of operation and application area of basic Dual current fed converter with switch capacitors

Power conversion	93–96%	
Co/XMFR	73–85%	
Stab/Coeff	2.6, 1.4, 0.6, 0.6	1.3 average
Max commutation frequency	700 kHz	Limited by C1 and C2. Above 400 kHz Even more increase cost of C1 and C2
Start-up method	Itself	
Reliability	High	Dosen capacitors on both side
Input voltage max	300 VDC	Limited by semiconductors 1200 V
Input voltage min	2–7 VDC	Below 7VDC increase cost
Input current	Max 1200 A	Limited by cost of implementation low voltage power stage, even more above 600 A, include transformer and Csw capacitors
Range of vin	1.8 times	More than 1.8 start to suffer efficiency or need more complicated control
Load range	0–100%	
Output current	Around up 500 A	Limited by cost of implementation secondary side of power stage, include transformer
Response time	Slow/middle	Limit by resonance inductor
Direction	Step-up or step-down	Bi-directional. Replace output diodes to active switches

6.9 Supplemental

The content of this chapter represents the original, creative work of the authors. We utilized OpenAI's ChatGPT to check grammar, spelling, and improve the clarity of certain portions of the text.

References

1. C. Adragna, LLC resonant converters: an overview of modeling, control and design methods and challenges. Found. Trends Electr. Energy Syst. **5**, 75 (2022). https://doi.org/10.1561/3100000029
2. H. Li, Z. Zhang, S. Wang, J. Tang, X. Ren, Q. Chen, A 300-kHz 6.6-kW SiC bidirectional LLC onboard charger. IEEE Trans. Ind. Electron. **7**(2) (2020)
3. M. Dong, C. Chen, Z. Xu, H. Chen, D. Han, J. Zhang, A high-efficiency bidirectional LLC resonant converter based on current-Doubler rectification circuit, in *IEEE 5th International Electrical and Energy Conference (CIEEC)*, (2022)
4. J.-h. Jung, J.-g. Kwon, Theoretical analysis and optimal design of LLC resonant converter, in *European Conference on Power Electronics and Applications*, (2007)
5. R. Beiranvand, B. Rashidian, M.R. Zolghadri, S.M.H. Alavi, A design procedure for optimizing the LLC resonant converter as a wide output range voltage source. IEEE Trans. Power Electron. **27**(8), 3749 (2012)

6. Half-Bridge LLC Resonant Converter Design Using FSFR-Series Fairchild Power Switch (FPS™) AN-4151. www.fairchildsemi.com
7. Hong Huang Designing an LLC Resonant Half-Bridge Power Converter Power Seminar topics and online power training modules. power.ti.com/seminars
8. LLC Converter Design Note. www.infinion.com
9. C.-H. Yang, T.-J. Liang, K.-H. Chen, J.-S. Li, J.-S. Lee, Loss analysis of half-bridge LLC resonant converter, in *1st International Future Energy Electronics Conference (IFEEC)*, (2013)
10. Q. Wang, X. Zhang, R. Burgos, D. Boroyevich, A. White, M. Kheraluwala, Design considerations for a high efficiency 3 kW LLC resonant DC/DC transformer, in *IEEE Energy Conversion Congress and Exposition (ECCE)*, (2015)
11. Q. Cao, Z. Li, B. Xue, H. Wang, Fixed frequency phase shift modulated LLC resonant converter adapted to ultra wide output voltage range, in *IEEE Applied Power Electronics Conference and Exposition (APEC)*, (2019)
12. S. Alatai, M. Salem, D. Ishak, A. Bughneda, M. Kamarol, D.N. Luta, Phase-shifted LLC resonant DC-DC converter for battery charging application, in *IEEE conference on energy conversion (CENCON)*, (2021)
13. Y. Wei, A. Mantooth, An LLC converter with fixed switching frequency operation for renewable energy applications, in *IEEE Transportation Electrification Conference & Expo (ITEC)*, (2021)
14. Y. Wei, H. Quanming Luo, A. Mantooth, Synchronous rectification for LLC resonant converter: an overview. IEEE Trans. Power Electron. **36**(6), 7264 (2021)
15. E.S. Glitz, J.-D. Hsu, M. Ordonez, Power loss estimation in LLC synchronous rectification using rectifier current equations. IEEE Trans. Ind. Electron. **67**(5), 3696 (2020)
16. W. Guo, K.(H.) Bai, A.T.J. Patterson, J. Kane, A novel soft starting strategy of an LLC resonant DC/DC converter for plug-in hybrid electric vehicles, in *Twenty-Eighth Annual IEEE Applied Power Electronics Conference and Exposition (APEC)*, (2013)
17. C. Fei, F.C. Lee, Q. Li, Digital implementation of soft start-up and short-circuit protection for high-frequency LLC converters with optimal trajectory control (OTC). IEEE Transactions on Power Electronics **32**, 32 (2017)
18. C. Fei, F.C. Lee, Q. Li, Soft start-up for high frequency LLC resonant converter with optimal trajectory control, in *IEEE Applied Power Electronics Conference and Exposition (APEC)*, (2015)
19. Y. Ling, Z. Guo, X. Liu, The soft-start analysis of a full-bridge LLC converter with hybrid control strategy, in *IEEE Vehicle Power and Propulsion Conference (VPPC)*, (2016)
20. A. Nabih, M.H. Ahmed, Q. Li, F.C. Lee, Transient control and soft start-up for 1-MHz LLC converter with wide input voltage range using simplified optimal trajectory control. IEEE J. Emerg. Select. Topics Power Electron. **9**, 24 (2022)
21. Q. Cao, Z. Li, H. Wang, Wide voltage gain range LLC DC/DC topologies: state-of-the-art, in *International Power Electronics Conference (IPEC-Niigata 2018-ECCE Asia)*, (2018)
22. H. Kato, S. Karasuyama, Y. Ishizuka, K. Ogasawara, K. Ueda, Optimization of a transformer used for LLC resonant DC-DC converters for photovoltaic cells, in *IEEE International Telecommunications Energy Conference (INTELEC)*, (2015)
23. Y. Wei, Q. Luo, A. Mantooth, Overview of modulation strategies for LLC resonant converter. IEEE Trans. Power Electron. **35**(10), 10423 (2020)
24. B. Mammano, J. Putsch, Phase-shift PWM control-a new integrated controller eases the design of efficient high frequency bridge switching, in *APEC1991*, (1991)
25. Application Note AN2013–03. 2013, Infineon
26. Phase-Shifted Full Bridge DC/DC Power Converter Design Guide. 2014, Texas Instrument
27. Application Note, *Current Doubler Rectifier Offers Ripple Current Cancellation* (Texas Instrument, 2004)
28. S. Zeljkovic, T. Reiter, D. Gerling, Analysis of rectifier topologies for automotive HV to LV phase shift ZVT DC/DC converter, in *EPE-PEMC 2012 ECCE Europe*, (2012)

29. P. Alou, J.A. Oliver, O. García, R. Prieto, J.A. Cobos, Comparison of current doubler rectifier and center tapped rectifier for low voltage applications, in *APEC 2006*, (2006)
30. M. Basler, R. Reiner, S. Moench, P. Waltereit, R. Quay, GaN active rectifier diode. IEEE Open J. Power Electron. (2022)
31. D. Reusch, J. Glaser, Design considerations for GaN transistor based synchronous rectification, in *APEC 2018*, (2018)
32. Application Note, 2022. 3-kW Phase-Shifted Full Bridge with Active Clamp Reference Design with > 270 W/in3 Power Density. Texas Instrument
33. J.-G. Cho, J.-W. Baek, C.-Y. Jeong, G.-H. Rim, Novel zero-voltage and zero-current-switching full-Bridge PWM converter using a simple auxiliary circuit. IEEE Trans. Ind. Appl. **35**(1) (1999)
34. P. Rehlaender, T. Grote, S. Tikhonov, M. Schroeder, F. Schafmeister, J. Boecker, A 3.6 kW single-stage LLC converter operating in half-bridge, full-bridge and phase-shift mode for automotive onboard DC-DC conversion, in *PCIM Europe Digital Days 2020*, (2020)
35. G.N.B. Yadav, N. Lakshmi Narasamma, An active soft switched phase-shifted full-Bridge DC–DC converter: analysis, modeling, design, and implementation. IEEE Trans. Power Electron. **33**(4), 4538 (2014)
36. J.-K. Han, G.-W. Moon, Circulating current-less phase-shifted full-Bridge converter with new rectifier structure, in *2018 International Power Electronics Conference (IPEC-Niigata 2018-ECCE Asia)*, (2018)
37. Power conversion apparatus for DC/DC conversion using Dual Active Bridges, Patent 5,027,264 US, 25 June 1991
38. R.W. DeDoncker, D.M. Divan, M.H. Kheraluwala, A three-phase soft-switched high power-density Dc/Dc converter for high-power applications, in *Conference IEEE Industry Applications*, (1988)
39. M.N. Kheraluwala, R.W. Gascoigne, D.M. Divan, E.D. Baumann, Performance characterization of a high-power dual active Bridge dc-to-dc converter. IEEE Trans. Ind. Appl. **28**(6), 1294 (1992)
40. N.H. Baars, J. Everts, H. Huisman, J.L. Duarte, E.A. Lomonova, A 80-kW isolated DC–DC converter for railway applications. IEEE Trans. Power Electron. **30**(12), 6639 (2015)
41. F. Krismer, J.W. Kolar, Efficiency-optimized high-current dual active bridge converter for automotive applications. Trans. Ind. Electron. **59**(7) (2012)
42. R.T. Naayagi, A.J. Forsyth, R. Shuttleworth, High-power bidirectional DC–DC converter for aerospace applications. IEEE Trans. Power Electron. **27**(11), 4366 (2012)
43. M. Ryu, D. Jung, J. Baek, H. Kim, An optimized design of bi-directional dual active bridge converter for low-voltage battery charger, in *16th International Power Electronics and Motion Control Conference and Exposition*, (2014)
44. R. Haneda, H. Akagi, Design and performance of the 850-V 100-kW16-kHz bidirectional isolated DC–DC converter using SiC-MOSFET/SBD H-Bridge modules. IEEE Trans. Power Electron. **35**(10), 10013 (2020)
45. A.K. Jain, R. Ayyanar, PWM control of dual active bridge: comprehensive analysis and experimental verification. IEEE Trans. Power Electron. **26**(4), 1215 (2011)
46. D. Sha, X. Guo, *High-Frequency Isolated Bidirectional Dual Active Bridge DC–DC Converters with Wide Voltage Gain* (Springer, 2019)
47. Z. Guo, *Deshang Sha New Topologies and Modulation Schemes for Soft-Switching Isolated DC–DC Converters* (Springer)
48. F. Krismer, PhD Dissertation, Modeling and optimization of bidirectional dual active bridge DCDC converter topologies, 2010
49. F. Krismer, J.W. Kolar, Accurate power loss model derivation of a high-current dual active Bridge converter for an automotive application. IEEE Trans. Ind. Electron. **57**(3), 881 (2010)
50. H. Li, F.Z. Peng, J.S. Lawler, A natural ZVS medium-power bidirectional DC-DC converter with minimum number of devices. IEEE Trans. Ind. Appl. **39**(2), 525 (2003)

51. H. Daneshpajooh, A. Bakhshai, P. Jain, Optimizing dual half bridge converter for full range soft switching and high efficiency, in *2011 IEEE Energy Conversion Congress and Exposition*, (2011)
52. A. Isurin, A. Cook, A Novel resonant converter topology and its application, in *IEEE PESC*, (2001), pp. 1039–1044
53. Alexander Topology Resonance Energy Conversion and Inversion Circuit Utilizing a Series Capacitance Multi-voltage Resonance Section, Patent 6,483,731 US, 19 Nov 2002
54. High-Frequency DC-DC Converter Control, Patent 7,379,309 US, 27 May 2008
55. A. Isurin, A. Cook, Cost effective resonant DC-DC converter for hi-power and wide load range operation, in *IEEE ISIE 2006*, (2006)
56. A. Isurin, A. Cook, *9kW Isolated DC-DC Converter for Hybrid Bus PCIM 2012*
57. A. Isurin, A. Cook, Achieving a wide input voltage and output power load range step-down DC-DC conversion with good full range efficiency at 4kW, in *APEC 2006*, (2006)
58. Y. Shen, H. Wang, Z. Shen, Y. Yang, F. Blaabjerg, A 1-MHz series resonant DC–DC converter with a dual-mode rectifier for PV microinverters. IEEE Trans. Power Electron. **34**(7), 6544 (2019)
59. Y. Shen, H. Wang, A. Al-Durra, Z. Qin, F. Blaabjerg, A bidirectional resonant DC–DC converter suitable for wide voltage gain range. IEEE Trans. Power Electron. **33**(4), 2957 (2018)
60. Y. Shen, H. Wang, Z. Qin, F. Blaabjerg, A. Al Durra, A reconfigurable series resonant DC-DC converter for wide-input and wide-output voltages, in *APEC 2017*,
61. C. Bai, M. Kim, Bidirectional resonant converter with minimized switching loss over wide operating voltage range. IEEE J. Emerg. Select. Topics Power Electron. **10**(3), 2975 (2022)
62. J.-M. Kwon, B.-H. Kwon, High step-up active-clamp converter with input-current Doubler and output-voltage Doubler for fuel cell power systems. IEEE Trans. Power Electron. **24**(1), 108 (2009)
63. M.C. Mira, Z. Zhang, A. Knott, M.A.E. Andersen, Analysis, design, modeling, and control of an interleaved-boost full-Bridge three-port converter for hybrid renewable energy systems. IEEE Trans. Power Electron. **32**(2) (2017)
64. H. Xiao, S. Xie, A ZVS bidirectional DC–DC converter with phase-shift plus PWM control scheme. IEEE Trans. Power Electron. **23**(2), 813 (2008)
65. Z. Yan, J. Zeng, J. Liu, W. Lin, A novel soft-switching bidirectional dc-dc converter with high voltage-gain for grid-connected energy storage system, in *4th IEEE Workshop on the Electronic Grid (eGRID)*, (2019)
66. Y. Li, Y. Xing, Y. Lu, H. Wu, P. Xu, Performance analysis of a current-fed bidirectional LLC resonant converter, in *IECON 2016—42nd Annual Conference of the IEEE Industrial Electronics Society*, (2019)
67. H. Wu, K. Sun, Y. Li, Y. Xing, Fixed-frequency PWM-controlled bidirectional current-fed soft-switching series-resonant converter for energy storage applications. IEEE Trans. Ind. Electron. **64**(8), 6190 (2017)
68. X. Pan, H. Li, Y. Liu, T. Zhao, J. Chenchen, A.K. Rathore, An overview and comprehensive comparative evaluation of current-fed-isolated-bidirectional DC/DC converter. IEEE Trans. Power Electron. **35**(3) (2020)
69. Z. Zhang, W. Zhiying, S. Xie, X. Ma, X. Jinming, M. Liu, A soft-switching current-fed isolated bidirectional DC-DC converter with low circulating power and easy-implemented control strategy, in *IEEE Energy Conversion Congress and Exposition (ECCE)*, (2019)
70. D. Sha, X. Yaxiong, J. Zhang, Y. Yan, Current-fed hybrid dual active Bridge DC–DC converter for a fuel cell power conditioning system with reduced input current ripple. IEEE Trans. Ind. Electron. **64**(8), 6628 (2017)
71. H. Seok, B. Han, B.-H. Kwon, M. Kim, High step-up resonant DC–DC converter with ripple-free input current for renewable energy systems. IEEE Trans. Ind. Electron. **65**(11), 8543 (2018)

72. J. Zeng, Z. Yan, J. Liu, Z. Huang, A high voltage-gain bidirectional DC–DC converter with full-range ZVS using decoupling control strategy. IEEE J. Emerg. Select. Topics Power Electron. **8**(3), 2775 (2020)
73. H. Kim, J. Park, S. Kim, R.M. Hakim, H. Belkamel, S. Choi, A single-stage electrolytic capacitor-less EV charger with single- and three-phase compatibility. IEEE Trans. Power Electron. **37**(6), 6780 (2022)
74. H. Belkamel, H. Kim, S. Choi, Totem-pole ZVS converter operating in CCM for single-stage bidirectional AC–DC conversion with high-frequency isolation. IEEE Trans. Power Electron. **36**(3), 3486 (2021)
75. A. Isurin, A. Cook, Dual fed, step up converter topology and its application, in *PCIM Europe 2015, Nuremberg, Germany*, (2015)
76. A. Isurin, A. Cook, Step-up DC-DC converter for automotive application, in *18th European Conference on Power Electronics and Applications (EPE'16 ECCE Europe)*, (2016)
77. F. Shang, M. Krishnamurthy, A. Isurin, A novel high gain step-up resonant DC-DC converter for automotive application, in *IEEE Applied Power Electronics Conference and Exposition (APEC)*, (2016)
78. F. Shang, G. Niu, M. Krishnamurthy, Design and analysis of a high-voltage-gain step-up resonant DC–DC converter for transportation applications. IEEE Trans. Transp. Electrif. **3**(1), 157 (2017)
79. F. Shang, W. Haowen, G. Niu, M. Krishnamurthy, A. Isurin, Dynamic analysis and control approach for a high-gain step-up converter for electrified transportation. IEEE Trans. Transp. Electrif. **3**(3), 656 (2017)
80. F. Shang, PhD Dissertation, High gain high efficiency resonant DC-DC converter, 2016
81. Isolated Step-Up Converter, Patent Application Publication US 2017/0310221, 26 Oc 2017
82. Isolated Step-Up Converter, Patent US 9,871,450B2, 16 Jan 2018
83. Bi-Directional Isolated DC-DC converter for Electrification of Transportation, Patent US 10,483,862B2, 19 Nov 2019
84. Bi-Directional Isolated DC-DC converter for Electrification of Transportation, Patent US 11,063,518, 13 July 2021

Chapter 7
Comparison of Isolated DC/DC Converters

The primary goal of this book is to provide practical guidance on the design of isolated DC–DC converters. In practice, this means that a design engineer receives the technical specifications for a power converter that they need to develop, along with cost targets based on the planned production volume per year. In the automotive industry, adhering to cost constraints is a strict requirement. This is the first step. After analyzing this information, the next step involves the power electronics design, which includes selecting the topology and power components to be used. The selection process requires choosing one topology from a list of options. This decision is crucial, as it significantly influences the overall success of the project. Having a list of comparative topologies is essential for making an informed choice. This principle is fundamental in engineering and proves very useful, which is why many publications provide comparisons of topologies based on various parameters. Some publications compare several topologies [1, 2], while others focus on three [3] or two [4]. A particularly good approach is to compare the application areas of a single topology [5] or a family of related topologies [6]. Generally, these comparisons are valuable and provide useful information, but they are often incomplete. Unfortunately, while some studies mention cost considerations [1, 4, 5], they often lack specific numbers and do not detail which parts of the converter contribute most significantly to the overall cost or why. Consequently, conclusions about the cost-effectiveness of a particular topology can be quite controversial [4]. Another important aspect is the technical specifications that the design engineer receives, which pertain to the application area. It's highly preferable to understand the capabilities of a topology concerning the application and operating environment. For example, in references [7–9], the authors use data from a single implementation for their comparisons. If they mention the application, it indirectly indicates where this topology can be used [7]. All of this information is useful, but it is not comprehensive because it does not provide a complete view of the capabilities, operation, and application areas. Understanding the full range of a topology's capabilities helps

A. Isurin, M. J. Scott, *Practical Design Considerations for Isolated DC-DC Converters*, https://doi.org/10.1007/978-3-031-94427-7_7

design engineers select the most suitable topology for the specific targets and requirements they have received.

It's also important to note that the operation area influences the application area. In other words, the application area is an extension of the operation area.

7.1 Comparison of Six Topologies

In the previous chapter, we discussed six basic topologies that are generally suitable for use in isolated DC–DC converters with output power exceeding 2–3 kW and currents on the low voltage side over 100–200 A. These topologies are: (1) resonant LLC converter; (2) phase-shift (PS) converter; (3) dual active bridge (DAB) converter; (4) resonant with PWM (R-PWM) converter; (5) dual current-fed with active clamp (DCAC) converter; and (6) dual current-fed with switched capacitors (DCSC) converter. The authors have also selected parameters that we will use to compare these topologies against each other. Before we begin comparing the topologies, it's **important to emphasize that the primary goal for a design engineer in practical considerations is to develop a cost-effective power converter with superior performance.** Therefore, the total cost of a product and its development, including after-sales costs, are the primary focus of engineering. Power electronics design begins with the selection of the appropriate topology and power components. The authors do not claim that all numbers are exact, as the real cost and performance of converters are influenced by many individual factors. However, the authors are confident that the proportional representation provides a realistic overview, and design engineers can use this data as a reference.

7.2 Comparison by Cost

Let's start the comparison with cost. Note that the estimations were made under the following assumptions: all topologies are used for the same application area, all have the same commutation frequency, and exhibit similar efficiency. Changes in these parameters can significantly impact the total cost, but the ratio between parts of the converter—such as the power stage, auxiliary circuits, and filters—will remain relatively unchanged. Table 7.1 shows the cost of power components for the topologies considered in the step-down mode, and Table 7.2 presents the cost of power components for the topologies considered in the step-up mode.

The cost of the control board, housekeeping power supply, and measuring circuits is not included in this analysis, as these costs are consistent across all topologies and do not affect the cost of the power section. Analyzing the data from these tables, we observe that the cost of the power stages themselves does not vary significantly. The primary difference in cost arises from the size of the magnetic components in the filters. This influence is illustrated by the cost variations in the LLC

Table 7.1 The cost of power parts/components of the topologies in the step-down mode

Topology	Total cost $/percentage	Power stage $/percentage	Aux circuits $/percentage	Filters $/percentage	Coefficient $/percentage
DAB	1109.04/100	422.94/38.1	116.16/10.5	187.06/16.9	369.68/34.5
PS	866.56/100	426.16/49.2	63.68/7.3	87.87/10.1	288.85/33.4
R-PWM	561.54/100	389.36/69.3	12.52/2.2	66.07/11.8	93.59/16.7
LLC/common	654.24/100	375.2/57.3	2/0.3	126.06/19.3	150.62/23.0
LLC/without	520.63/100	375.2/72.1	2/0.4	56.66/10.9	86.53/16.6

Table 7.2 The cost of power parts/components of the topologies in the step-up mode

Topology	Total cost $/percentage	Power stage $/percentage	Aux circuits $/percentage	Filters $/percentage	Coefficient $/percentage
DAB	1109.04/100	422.94/38.1	116.16/10.5	187.06/16.9	369.68/34.5
R-PWM	468.76/100	312.04/66.6	12.52/2.7	66.07/14.1	78.13/16.6
DCAC	680.47/100	349.72/51.4	26.72/3.9	109.61/16.1	194.42/28.6
DCSC/ceramic	643.03/100	439.89/68.4	N/A	54.75/8.5	148.39/23
DCSC/film	531.63/100	339.05/63.8	N/A	54.75/10.3	137.83/25.9

topology. It's important to note that changes in filter costs are related to changes in the application area.

The key cost driver is the expense of passive components, as evidenced by the cost differences between two implementations of the DCSC topology. In this case, there are no changes in the operating area, only in the size of the converter. The impact of size reduction on cost is a separate topic for discussion. Additionally, the cost of filters indirectly reflects the level of EMI associated with each topology. In other words, higher costs typically indicate higher EMI, while lower costs suggest lower EMI.

Auxiliary circuits also have a significant impact on the total cost of the converter, as seen in the PS topology and more prominently in the DAB topology. The conclusion of this section is that, for a cost-effective design, one should select a topology with natural commutation of active switches and minimal EMI. Soft switching technology is particularly beneficial for achieving a cost-effective product. Furthermore, costs are reduced when the output rectifier does not use synchronous rectification, as illustrated by the R-PWM in the step-up mode.

7.3 Comparison by the Operation Area

The next point is the capability of the operational areas of the basic topologies under consideration. Table 7.3 shows the capability of these operational areas.

Analyzing the data, we observe that pure resonant topologies are more reliable and can operate at higher commutation frequencies compared to non-resonant topologies, such as DAB and PS. Note that the maximum commutation frequency is

Table 7.3 The capability of the operation area of tested topologies

Topology	P-C %	Co-Tr %	Stab Aver	Soft switch	Soft-start	Relay	Freq max kHz
DAB	78–91	70–90	0.4/0.6	Narrow	Complicated	Low	100 kHz
PS	60–85	69–85	0.81	Need aux circuits	Itself with snubber	Middle	200 kHz
R-PWM Step-down	85–96	87–92	0.97	Whole range Itself	Itself	High	1000 kHz
R-PWM Step-up	93–95	84–87	0.97	Whole range Itself	Itself or special control	Middle High	1000 kHz
LLC	56–89	52–62	0.957	Whole range Itself	Special control	Middle	1000 kHz
DCAC	77–92	86–92	0.77	Whole range With snubber	Itself	Middle High	200 kHz
DCSC	93–96	73–85	1.3	Whole range Itself	Itself	High	700 kHz

determined by the resonance tank. Generally, higher commutation frequencies help reduce the size of filters and associated costs. Another important point is that topologies with variable frequency control under full power, such as LLC, have a significantly smaller number of power conversion coefficients (P-C) and coefficients using power transformer (Co-XMFR). This means that with this topology and transformer, design engineers face additional challenges in optimizing cost and size compared to other topologies with fixed commutation frequencies under full power load. We will now compare the capabilities of these topologies by application area. The first point of comparison is the maximum power each topology can convert, determined by the maximum input current the topology can accept without damage. The maximum input voltage is limited by the maximum voltage that the semiconductors in the input section can handle. By multiplying the maximum input voltage (Vin-max) by the maximum input current (Iin-max), we can determine the maximum power that the topology can convert. A similar consideration applies to the output power. The maximum output power of a converter is determined by the maximum current it can produce. By multiplying the maximum output voltage (Vout-max) by the maximum output current (Iout-max), we can calculate the maximum power the converter can deliver to the load. In other words, the smaller of these values represents the maximum power the converter can produce under the conditions used for these calculations. This comparison is divided into two different sections: step-down and step-up. Each section is further divided into low-cost and high-cost categories, as the specifications of the application area influence the cost.

7.4 Comparison by Application Area

Tables 7.4 and 7.5 show the application areas for the step-down mode, while Tables 7.6 and 7.7 illustrate the application areas for the step-up mode.

Table 7.4 The capability of application area in step-down by parameters of input and output with maximum possibility

Topology	Vin-max	Vin-min	Iin-max	Vin range	Load range	Iout max	Response time
DAB	750 V	30 V	100A	1.3	60–100%	100A	Slow
PS	750 V	30 V	100A	1.8	30–100%	600A	Slow
R-PWM	850 V	30 V	80A	3	0–100%	1000A	Fast
LLC	850 V	30 V	100A	1.5	0–100%	1000A	Middle

Table 7.5 The capability of application area in step-down by parameters of input and output with cost effective range

Topology	Vin-max	Vin-min	Iin-max	Vin range	Load range	Iout max	Response time
DAB	750 V	100 V	100A	1.3	60–100%	100A	Slow
PS	750 V	100 V	100A	1.6	30–100%	300A	Slow
R-PWM	850 V	200 V	50A	2	0–100%	500A	Fast
LLC	850 V	200 V	50A	1.25–1.3	0–100%	500A	Middle

Table 7.6 The capability of application area in step-up by parameters of input range and output load range with maximum possibility

Topology	Vin-max	Vin-min	Iin-max	Vin range	Load range	Iout max	Response time
DAB	750 V	30 V	100A	1.3	60–100%	100A	Slow
R-PWM	850 V	10 V	700A	2	0–100%	80A	Fast
DCAC	400 V	2 V	400A	1.8	0–100%	500A	Fast-middle
DCSC	300 V	2 V	1200A	1.8	0–100%	500A	Middle-slow

Table 7.7 The capability of application area in step-up by parameters of input range and output load range with cost effective range

Topology	Vin-max	Vin-min	Iin-max	Vin range	Load range	Iout max	Response time
DAB	750 V	30 V	100A	1.3	60–100%	100A	Slow
R-PWM	850 V	30 V	500A	1.8	0–100%	50A	Fast
DCAC	400 V	7 V	400A	1.8	0–100%	500A	Fast-middle
DCSC	300 V	7 V	600A	1.8	0–100%	500A	Middle-slow

As we can see, the main difference between the maximum capability tables and the cost-effective tables is the change in current values and input voltage values. Additionally, the basic Dual Active Bridge (DAB) topology is not suitable for real implementation at currents over 100A. In other words, while the basic DAB topology may appear appealing, it requires significant effort to create an effective converter in practical applications. The main conclusion from analyzing the application area is that cost significantly influences the application area, and vice versa; they are interconnected. The cost differences (as shown in Table 7.1) for converters with LLC topology illustrate the potential for cost reduction by narrowing the application area. Furthermore, for a fast response time to large load changes, it is necessary to use a topology with a fixed commutation frequency and only a resonance tank, without any additional inductors.

7.5 Conclusion

The main conclusion from comparing the tested basic topologies is that for isolated DC–DC converters with currents ranging from 100 to 200 ADC and above, and for voltages below 200 VDC, one of the following three approaches is necessary: splitting the current, implementing ZCS, or employing both methods. These are strategic directions for the low-voltage side. On the high-voltage side, it is important to use both ZCS and ZVS with variable switch voltage. For higher currents, lower commutation voltages are required, while for lower currents, higher voltages are necessary. This is demonstrated by the use of resonant-pulse width Modulation (R-PWM) and dual-current source converter (DCSC) technologies. In other words, the use of soft-switching technology is essential. Note that implementing soft-switching technology requires a higher level of engineering expertise, but the final product will be more reliable and cost-effective compared to using hard-switching technology.

7.6 Supplemental

The content of this chapter represents the original and creative work of the authors. We utilized OpenAI's ChatGPT to check grammar, spelling, and improve the clarity of certain portions of the text.

References

1. M. Forouzesh, Y.P. Siwakoti, S.A. Gorji, F. Blaabjerg, B. Lehman, Step-up DC–DC converters: a comprehensive review of voltage-boosting techniques, topologies, and applications. IEEE Trans. Power Electron. **32**(12), 9143 (2017)
2. F. Alaql, I. Batarseh, Review and comparison of resonant DC-DC converters for wide-input voltage range applications, in *2019 IEEE Conference on Power Electronics and Renewable Energy (CPERE)*, (2019)
3. X. Li, Y. Zhang, P. Fang, J. Liu, Comprehensive comparison of three typical bridge structure isolated soft switching DC-DC topologies in the application of locomotive traction, in *2019 IEEE International Conference on Industrial Technology (ICIT)*, (2019)
4. T.M. Patarau, D.M. Petreus, I. Ferencz, Z. Orban, Comparison between LLC and phase-shift converter with synchronous rectification for high power, high current applications, in *2020 IEEE 26th International Symposium for Design and Technology in Electronic Packaging (SIITME)*, (2020)
5. J. Shahsevani, R. Beiranvand, *An Application-Oriented Review of the LLC-Based Resonant Converters*, vol 12 (IEEE Access, 2024), p. 52687
6. L.F. Costa, G. Buticchi, M. Liserre, A family of series-resonant DC–DC converter with fault-tolerance capability. IEEE Trans. Ind. Appl. **54**(1), 335 (2018)

7. L. Anbazhagan, J. Ramiah, V. Krishnaswamy, D.N. Jayachandran, A comprehensive review on bidirectional traction converter for electric vehicles. Int. J. Electron. Telecommun. **65**(4), 635 (2019)
8. K. Bhatt, R.A. Gupta, N. Gupta, Comparative analysis and control of bidirectional DC-DC converters, in *IEEE 1st International Conference on Energy, Systems and Information Processing (ICESIP)*, (2019)
9. G. Lithesh, B. Krishna, V. Karthikeyan, Review and comparative study of bi-directional DC-DC converters, in *IEEE International Power and Renewable Energy Conference (IPRECON)*, (2012)

Chapter 8
Power Magnetics

To implement a real-world isolated DC–DC converter, various components are required. The main units include the power stage, filter, control, and auxiliary circuits. Our focus is on the power stage, which comprises semiconductors, capacitors, resistors, magnetic components, a power transformer, and power inductors. While semiconductors, capacitors, and resistors can generally be purchased off the shelf, magnetic components often require custom design, especially when handling high currents exceeding 100–200 A.

8.1 Power Transformer

The power transformer is a crucial component of the power stage in an isolated DC–DC converter. It provides galvanic isolation and adjusts voltage levels to match the input and output requirements. This component is also relatively expensive, typically accounting for about 20–35% of the total cost of the power stage (in 2025). Due to its importance and cost, we will focus on the power transformer first.

In transportation applications, DC–DC converters commonly have specifications that include a low-voltage side ranging from 12 to 48 VDC with currents between 100 and 600 A, and a high-voltage side ranging from 270 to 800 VDC. Designing a high-frequency transformer to accommodate a 12–48 V winding and a high current of 500 A presents a significant challenge, as it often requires just a single turn in the winding. This situation poses substantial design challenges for engineers.

Another critical consideration is the step-up mode in DC–DC converters. Step-up conversion is generally more challenging than step-down, especially when the input voltage is 30 VDC or below. The main challenge arises from the power transformer. A transformer with two turns typically performs better than one with a single turn, due to better coupling, reduced magnetic field distortion, and more efficient core excitation.

A. Isurin, M. J. Scott, *Practical Design Considerations for Isolated DC-DC Converters*, https://doi.org/10.1007/978-3-031-94427-7_8

A crucial point is that the duty cycle of the power conversion should not fall below 90% (considering a half-cycle of power conversion), and the conversion process should be unidirectional. Otherwise, efficiency can significantly decrease, and costs may increase. This discussion focuses on high-frequency power transformers operating in the tens of kHz and higher, with power ratings exceeding 2–3 kW and handling currents of over 100–200 A on the low-voltage side.

There are many books and articles available on transformers, ranging from theoretical works [1] to practical guides [2, 3], as well as numerous articles [4]. While methodologies like those in [5] and others are commonly used, they often overlook transformers designed for high current and low-voltage applications, especially in practical implementations. In this discussion, we will focus on the practical aspects of designing high-frequency power transformers capable of handling currents over 200 A, with an emphasis on cost-effective solutions.

Figure 8.1 shows the equivalent schematic of a transformer with parasitic elements [2]. In this diagram:

- **R1** represents the resistance of the primary winding.
- **L1** is the primary leakage inductance.
- **Lm** is the magnetizing inductance.
- **Rc** represents the core loss resistance.
- **R2** is the resistance of the secondary winding.
- **L2** is the secondary leakage inductance.
- **C1** denotes the capacitance of the primary winding.
- **C2** denotes the capacitance of the secondary winding.
- **Cw** represents the interwinding capacitance.

The values of R1, R2, and Rc are crucial as they determine the power losses in the transformer, which we will discuss in more detail later. The leakage inductance, characterized by L1 and L2, can be a useful parameter, especially when the transformer operates in a resonant topology. In such cases, the leakage inductance can either contribute to or entirely function as the resonant inductor.

The magnetizing inductance, Lm, and the associated magnetizing current can be utilized to provide commutation current. This parameter must have a repeatable and

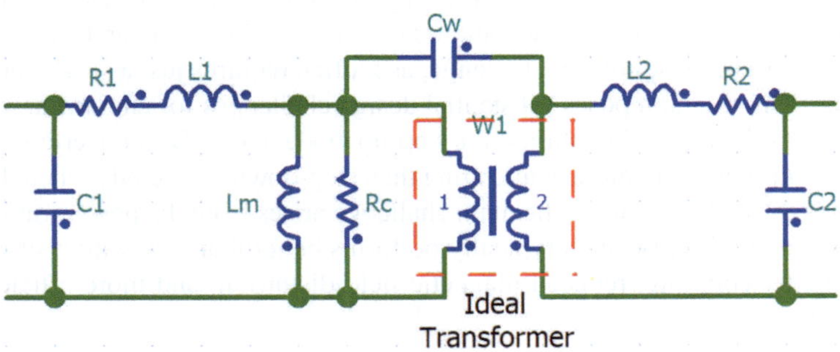

Fig. 8.1 The equivalent schematic of transformer with parasitic elements

precise value, necessitating the inclusion of an air gap in the power transformer. The authors recommend using a small air gap of around 0.1 mm, which helps prevent transformer saturation during short transient periods.

Transformer capacitance introduces additional considerations. These capacitances charge and discharge with each half cycle, leading to voltage spikes that generate electromagnetic interference (EMI) [6, 7]. Particularly concerning is the interwinding capacitance, which allows signals to pass between the windings, effectively reducing the isolation and contributing to EMI. This reduction in isolation can significantly impact the reliability of the power converter and the stability of the control loop.

Unfortunately, as commutation frequency and the slew rate of voltage change (dV/dt) increase, the issues associated with transformer capacitance also escalate. While it's impossible to eliminate these capacitances, efforts can be made to minimize them as much as possible [8, 9]. Currently, wide bandgap (WBG) semiconductors offer increasingly attractive options for use in power electronics. These semiconductors enable higher commutation frequencies, which in turn reduce the size and weight of components, thereby increasing power density. This parameter is especially crucial in the transportation sector.

However, interwinding capacitance poses a significant challenge by generating interference, which complicates the increase of commutation frequencies in high-power transformers. Now, let's consider the resistance of the windings and the core. As mentioned earlier, these resistances are key determinants of power losses in the transformer, directly affecting its efficiency. The primary source of power losses is the resistive (conductive) losses in the windings.

Table 8.1 shows the power losses of different power transformers, which were used by authors. It shows that conductive losses are mostly several times higher than core losses. These measurements are only by the transformer itself, without connections to the PCB, parts of the PCB, and busbars. We can use these data more or less for comparison, as a reference point.

In practice, a power transformer cannot be used in isolation; it is a crucial component of the power converter system. Similarly, a power stage without a power transformer is not functional. Therefore, it is essential for design engineers to consider the entire power converter system as a whole.

Table 8.1 The power losses of different power transformers

	Power	Ratio	Frequency	Cond	Core	Efficiency
U-core	64 kW	10–4	200 kHz	259 W	94 W	99.5%
	64 kW	10–4	100 kHz	147 W	134 W	99.6%
U-core	24 kW	13–1	200 kHz	225 W	29 W	98.9%
	24 kW	13–1	100 kHz	197 W	50 W	99.0%
U-core	15 kW	11–1	100 kHz	137 W	12 W	99.0%
Planar	7.5 kW	8–1	200 kHz	137 W	5.5 W	98.1%
	7.5 kW	8–1	120 kHz	112 W	5.5 W	98.4%

This means that the power transformer must be fully integrated into the power stage, both electrically and mechanically. The authors developed a method for this integration [10], with Fig. 8.2 illustrating the implementation of this idea in the first prototype for a 14 VDC, 250 A system. Although the idea has been modified multiple times since, the fundamental principle remains unchanged. Further details on integration implementation can be found in [11, 12].

Integrating the power transformer, especially on the low-voltage side, presents a significantly different scenario. Figure 8.3 shows a simplified schematic of the model used to analyze the interaction between the power transformer and the power stage. In this model:

- **L1** represents the total inductance of the transformer and its high-voltage leads (taps).
- **R1** denotes the total AC resistance of the transformer and its leads.
- **L2** indicates the total inductance of the connections and PCB.
- **R2** represents the total resistance of the connections and PCB.

The inductance observed on the high-voltage side, when the low-voltage side is shorted, is several times higher than the leakage inductance of the power transformer. This inductance is given by:

$$L\text{meas} = \text{L1} + \left(\text{L2} \times N^2\right), \tag{8.1}$$

where *L*meas is the measured inductance, and *N* is the turn ratio of the power transformer.

The current through L2 and R2 is substantial and exhibits a complex waveform, alternating between AC before the rectifier and DC with AC components after the rectifier. Conductive losses in the circuit comprising L2 and R2 can be significant

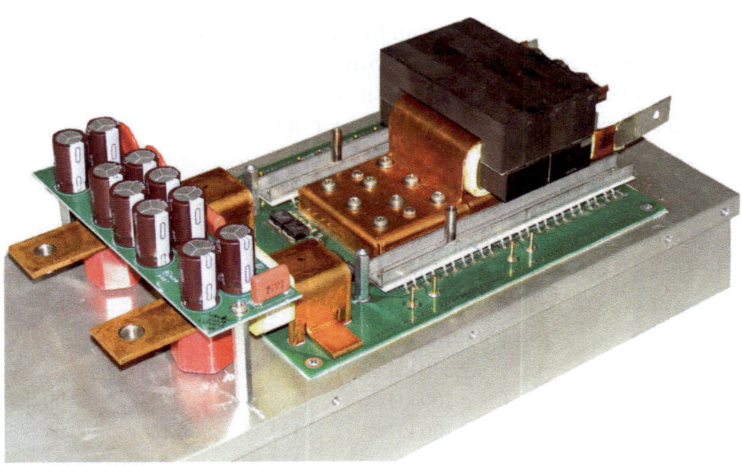

Fig. 8.2 The first prototype of patent US 7,123,123

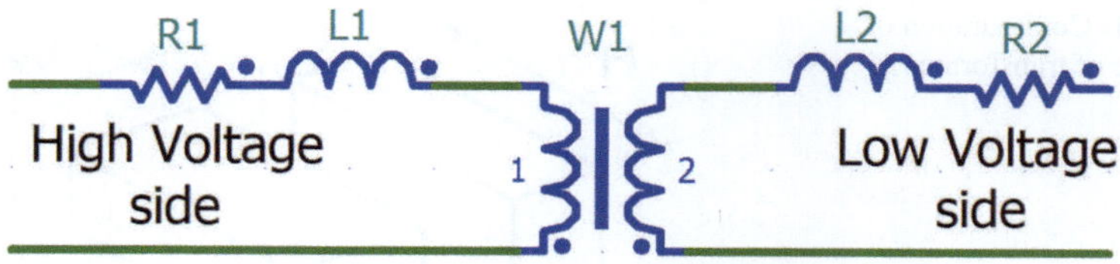

Fig. 8.3 Model of an ideal transformer and its connection and PCB

and depend on the commutation frequency. For a power stage with an output of 30 VDC, 500 A, operating at 100 kHz, and with 570 A AC current through the low-voltage winding of the transformer, the conductive losses can reach around 300 W. Including transformer losses and rectifier losses, the total losses amount to approximately 600–750 W.

While these losses are considerable, the transformer, L2, R2, and the rectifier benefit from excellent cooling, as they are mounted directly on the PCB, which in turn is attached to a heat sink. It is important to remember that the primary goals are low cost and superior performance, rather than efficiency alone.

The next step involves examining how to implement and optimize this circuit. For guidance on designing high-current and high-power transformers, refer to patent US 7,123,123, which outlines relevant design considerations.

As previously discussed, the main power losses on the lower-voltage side of the power transformer include conductive losses within the transformer itself and the conductive losses associated with connections between the transformer and the power stage. For high-voltage windings, LITZ wire is optimal, but attention must be given to the connections between the winding and the PCB. Standard ring terminals can have high resistance at high frequencies, potentially causing power losses of 2–4 W or more, which requires careful consideration.

The low-voltage side with a single turn can be implemented using a copper foil. When considering a single-turn winding for the low-voltage side and accommodating high current, attention must be paid to the ratio of the dimensions of the core's power leg, specifically the height (H) to width (W). This ratio is critical for optimizing the core's performance under high-current conditions.

Figure 8.4 illustrates the dimensions H (height) and W (width) of the power leg of a transformer. For most transformers, the H/W ratio typically falls between 1 and 2. However, for a single-turn transformer, this ratio should be maximized, ranging from 4 to 10, to ensure the foil winding is as long as possible inside the magnetic system. The foil winding inside the core window is functional and contributes to the transformer's performance. In contrast, the section of the foil winding outside the core window is primarily for connection purposes and can introduce additional conductive losses and inductance. To minimize these issues, the part of the foil winding outside the core should be kept as short as possible, serving mainly to connect the transformer to the PCB. Additionally, the contact area between the foil and the core acts as a cooling path for the foil winding. Therefore, a larger contact area enhances cooling efficiency.

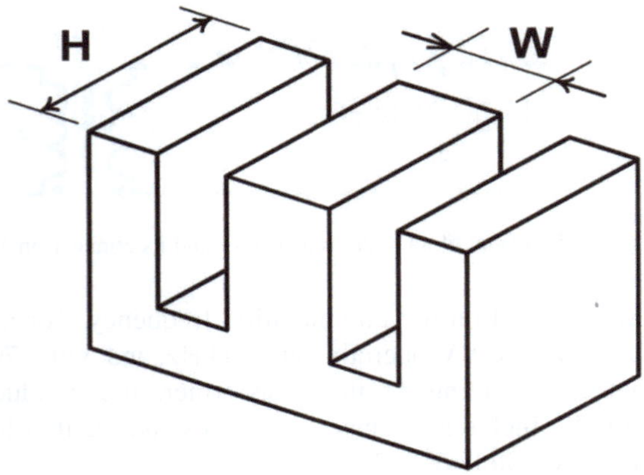

Fig. 8.4 Configuration of the core of transformer

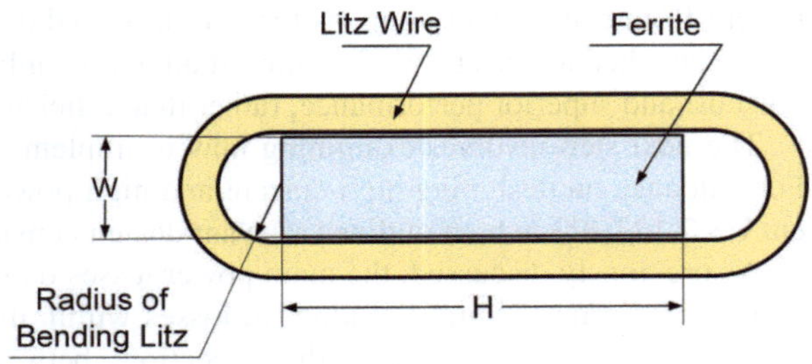

Fig. 8.5 The relationship between *W* and *H* for a magnetic core

Next, we discuss the power leg of the core, which is used for windings. The cross-sectional area of this leg is calculated based on the maximum flux density using the formula:

$$S_s = V * 10^4 / 4 * N * f * B \tag{8.2}$$

where S_s is the cross-sectional area, V is the voltage, N is the number of turns, f is the frequency, and B is the flux density. As mentioned earlier, the preferable *H/W* ratio should be as high as possible. Special consideration must be given to the minimum bending radius of LITZ wires, which will influence the size of the width (*W*) of the power leg.

Figure 8.5 illustrates the relationship between *W* and *H* for the power leg. The height (*H*) of the leg can be determined using the following formula:

$$H = S_s / W. \tag{8.3}$$

Next, we will consider another parameter of the power transformer: the cross-sectional area of the core window. The window of the transformer should be as wide as possible. This design choice offers two main benefits: a reduction in leakage inductance and an increase in the cross-sectional area of the foil, which helps lower the current density in high-current windings. It is preferable for the turns of LITZ wire to have gaps between each turn, using a single layer and positioning them on the bobbin. Figure 8.6 illustrates a possible arrangement of windings within the core's window. Additionally, the width of the foil determines the cross-sectional area available for connecting the winding to the PCB.

For the winding/tap connection of the foil to the PCB, using bolts with insulating shoulder washers is strongly preferred. While soldering can be used, it typically results in higher contact resistance compared to bolts with a pressure plate. Figure 8.7 demonstrates one possible method for connecting the winding to the PCB, where the width of the foil matches the width of the connection. The electrical connection is established through pressure, utilizing bolts with insulating shoulder washers and an L-shaped metal bar.

The thickness of the foil in the transformer is primarily determined by the skin effect. If the foil thickness exceeds the value calculated based on the skin effect, the AC resistance (Rac) will remain nearly the same, as indicated in Dowell's plots (Fig. 8.8 in [3]). However, using a thicker foil improves thermal conductivity because the increased mass of the foil enhances the transformer's cooling capability. Consequently, the thickness of the foil is often determined more by mechanical considerations than by electrical ones, typically ranging from 0.5 to 0.8 mm, though this can vary based on specific requirements.

We also considered the shapes and cross-sectional areas of the power leg and the window of the core. Power transformers can be constructed using two types of core shapes: E-core or U-core.

Figures 8.9, 8.10, and 8.11 illustrate possible implementations of these core types. Figure 8.9 shows a configuration where a single heatsink is used for both the power transformer and the low-voltage side of the power stage. This configuration is useful for topologies with a tap, where the primary objective of the pedestal is to provide a connection for the winding to the PCB and contribute to cooling. Additionally, the low-voltage side winding features a half-turn, requiring careful attention to ensure both windings are as equal as possible. Figure 8.10 depicts a configuration with two heatsinks: one for the power transformer and another for the low-voltage side of the power stage. Figure 8.11 illustrates a configuration with a U-core.

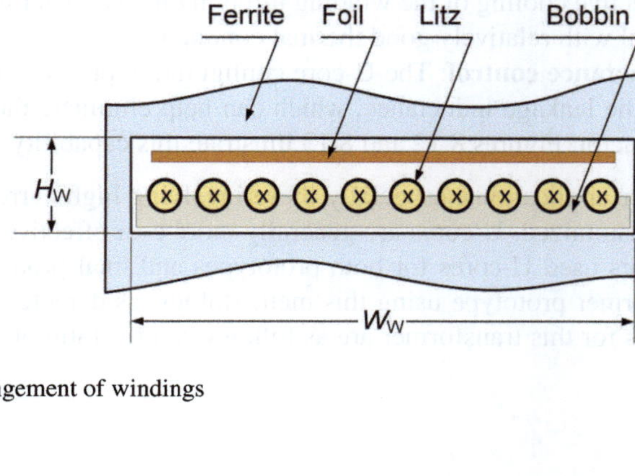

Fig. 8.6 Arrangement of windings

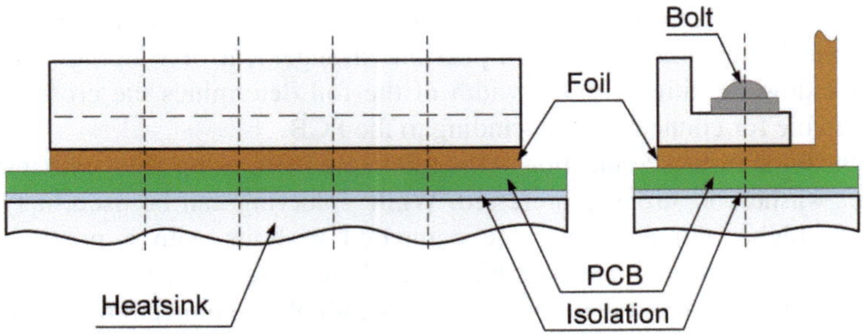

Fig. 8.7 Method for connecting the winding to the PCB

Fig. 8.8 The plot of Dowell

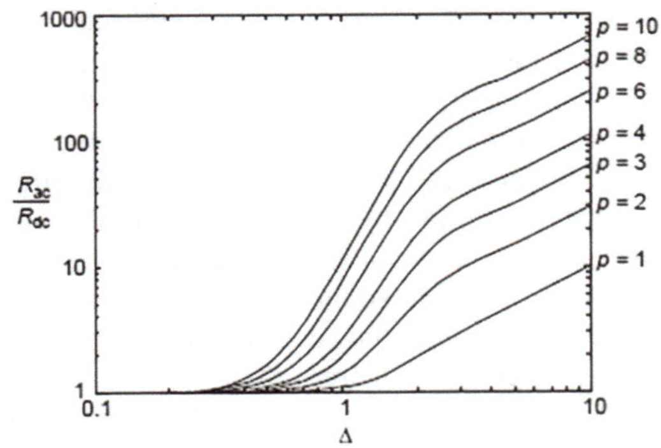

Let's compare the E-core and U-core configurations for transformers with a single turn. There are several key considerations:

1. **Length of the turn**: For a transformer with the same cross-sectional area in the power leg, the length of a single turn with AC current is shorter in a U-core configuration compared to an E-core.
2. **Cooling efficiency**: Transformers with a U-core offer better cooling for the winding compared to an E-core design. In a U-core transformer, the winding is placed directly on the PCB, which is mounted on a heat sink. This arrangement allows for effective cooling of the winding through direct contact with the PCB, a solid material with relatively good thermal conductivity.
3. **Leakage inductance control**: The U-core configuration provides more options for adjusting the leakage inductance, which can help eliminate the need for an additional inductor. Figures 8.12 and 8.13 illustrate this capability.

For high-power applications, especially those involving high currents, the core design is often customized. U-cores are generally more cost-effective in these scenarios. The authors used U-cores for both prototypes and final products. We constructed a transformer prototype using this methodology, as depicted in Fig. 8.13. The specifications for this transformer are as follows: a turns ratio of 11, operating

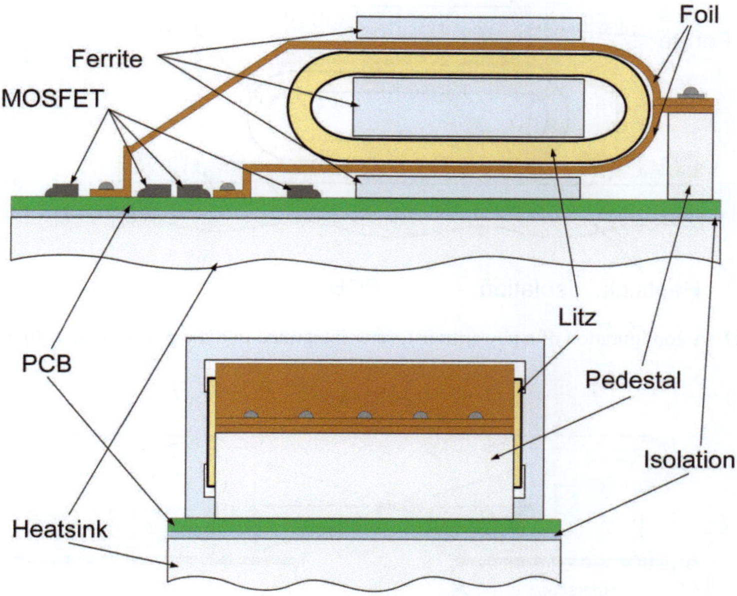

Fig. 8.9 An implementation where a single heatsink is used for both the power transformer and the low-voltage side of the power stage

Fig. 8.10 A winding configuration with two heatsinks

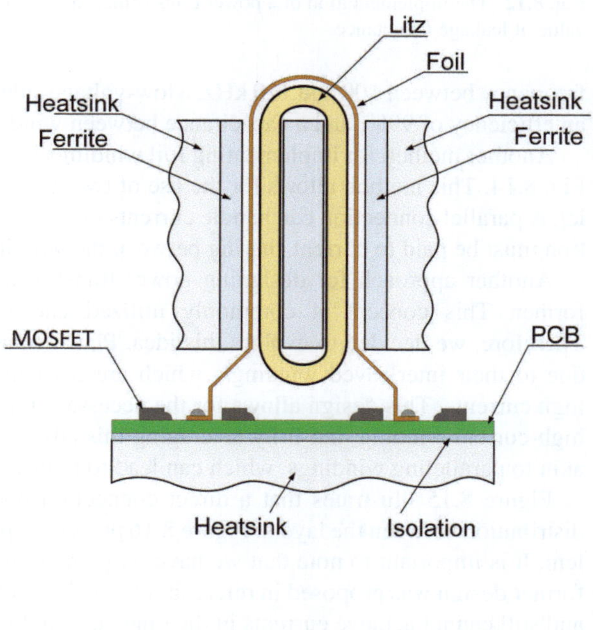

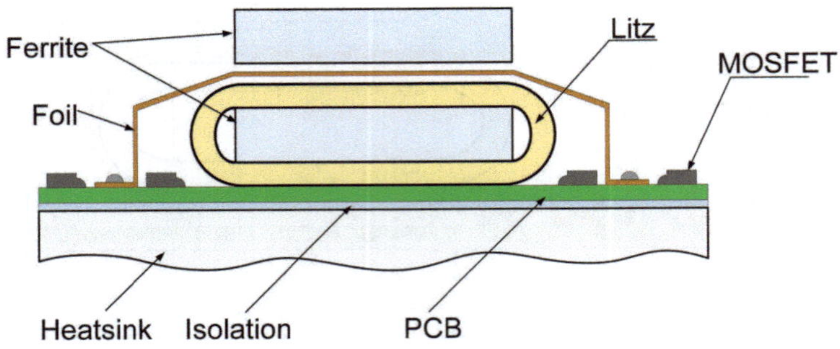

Fig. 8.11 A configuration of a power transformer integrated into the power stage with a U-core

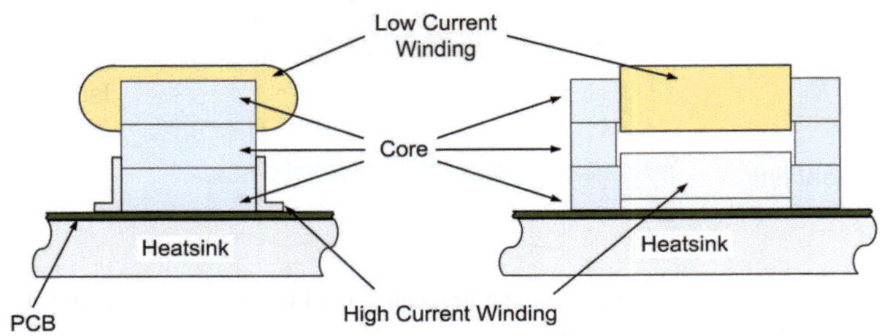

Fig. 8.12 The implementation of a power transformer integrated into the power stage with a high value of leakage inductance

frequency between 100 and 120 kHz, a low-voltage side current of 550 A at 30 VAC, an efficiency of 99%, and a capacitance between windings of 50 pF.

Another method for implementing foil windings is described in [13], as shown in Fig. 8.14. This method allows for the use of two turns connected in series or parallel. A parallel connection can handle currents of up to 800–900 A, but careful attention must be paid to current sharing between the windings.

Another approach for designing power transformers is to use a planar transformer. This concept is commonly utilized and initially appears promising. Therefore, we decided to explore this idea. Planar transformers have an advantage due to their interleaved windings, which are particularly beneficial for handling high currents. This design allows for the necessary total cross-sectional area of the high-current winding, but fully leveraging this advantage can be challenging. It is akin to paralleling windings, which can lead to issues with current sharing.

Figure 8.15 illustrates that a direct connection does not ensure equal current distribution between the layers. Figure 8.16 presents a possible solution to this problem. It is important to note that we have only one turn of winding. A matrix transformer design was proposed in references [14–16], but this approach is too complex and still cannot achieve currents in the range of hundreds of amperes.

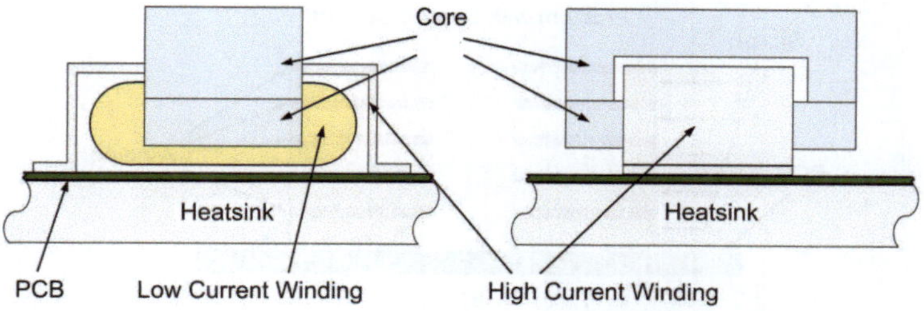

Fig. 8.13 The implementation of a power transformer integrated into the power stage with a lower value of leakage inductance

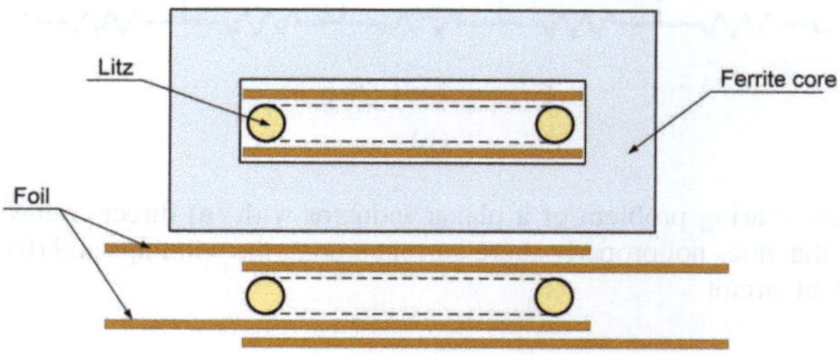

Fig. 8.14 Two-turn foil winding [13]

The following disadvantages significantly limit the applicability of planar transformers. Firstly, it is challenging to connect the high-current terminal of a planar transformer to a PCB, as the contact area between the transformer tap and the PCB is typically too small. This results in excessive heat generation at the contact point, which further elevates the temperature inside the transformer.

The authors compared a planar transformer with an integrated transformer, both used in a 9 kW DC–DC converter. The comparison was conducted using the same topology (Fig. 8.17) under the following conditions: input voltage (*V*in) of 600 V DC, output voltage (*V*out) of 30 V DC, and output current of 250 A DC (equating to 7.5 kW). The switching frequency was 125 kHz. Figure 8.18 provides images of these transformers.

The primary power losses in the planar transformer were concentrated in the contact area, with losses around 120 W. The small size of the contact area led to a high temperature of 175 °C after just 10 min of operation, indicating that the transformer requires a more complex cooling system. In such cases, engineers often resort to special liquid cooling for the contact area to manage the heat dissipation effectively.

Excessive interleaving of windings in a transformer can result in a high level of capacitance between the windings. This increased capacitance exacerbates electromagnetic interference (EMI) issues. The integrated transformer features a custom core, yet it is approximately 20% less expensive than the planar transformer. This cost difference and other comparative details are outlined in Table 8.2.

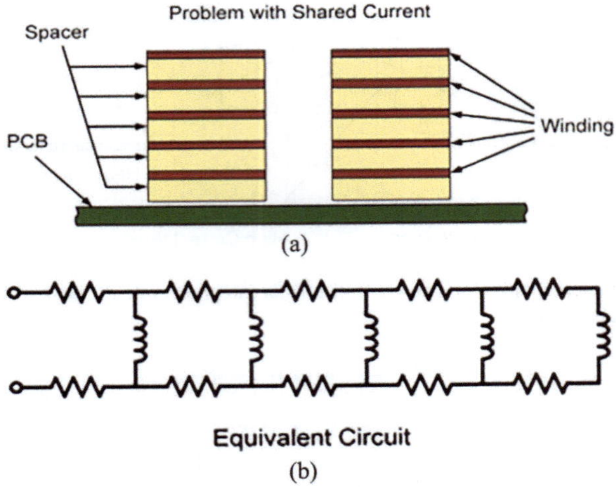

Fig. 8.15 Current sharing problem of a planar inductor with (**a**) direct connection of the low-voltage winding that does not properly share current among the windings and (**b**) the planar transformer's equivalent circuit

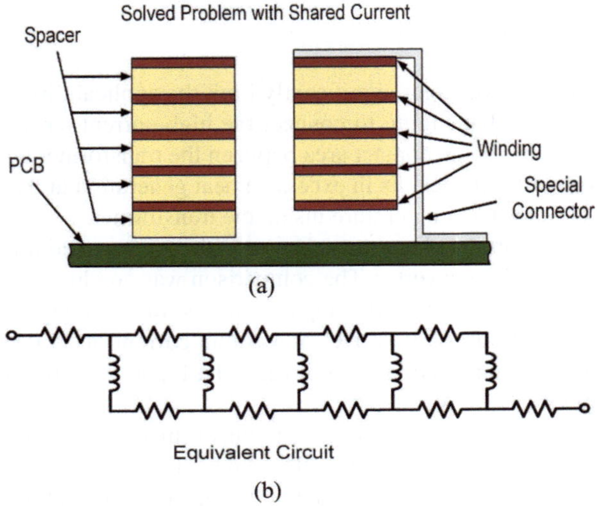

Fig. 8.16 Solution to the current sharing problem in a planar transform with (**a**) direct connection to the low-voltage winding of a planer transformer that properly shares current between all of the windings and (**b**) the planar transformer's equivalent circuit

The leakage inductance of planar transformers is a debated parameter. When using an E-core, the leakage inductance of planar transformers is not negligible [17]. Additionally, high-power planar transformers with E-cores occupy a large footprint, which can complicate mounting. Currently, ER-cores are more commonly

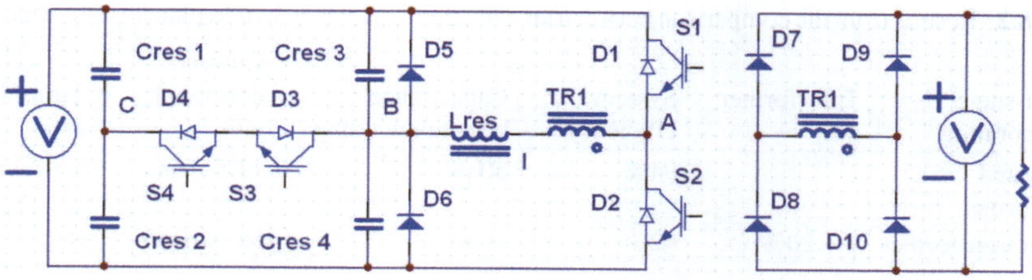

Fig. 8.17 Schematic of the Resonances with PWM topology that was used for the comparison in this section

Fig. 8.18 Photos of (**a**) the planar transformer and (**b**) the integrated transformers that were used for comparison. The (**b**) integrated transformer in the physical implementation of Fig. 8.12

used for high-power applications. ER-cores not only reduce mounting issues but also significantly decrease leakage inductance. However, this reduction in leakage inductance often necessitates the use of a larger additional inductor in the resonant tank. This requirement can increase the overall cost of the assembly, including the transformer and the resonant inductor.

8.1.1 Conclusions for Planar Transformers

8.1.1.1 Recommendations for Cost-Effective Planar Transformers

- **Power handling**: Best suited for applications not exceeding 1.5–2 kW.
- **Winding current**: Recommended for currents less than 50 A.
- **Isolation**: Suitable for winding isolation voltages below 1500 V.

Table 8.2 Results from the comparison between an integrated transformer and a planar transformer

Power stage incorporating	Transformer cost	System efficiency	Capacitance between winding	Current density of taps	Turns ratio
Integrated transformer	1	94%	50 pF	11.5 A/mm^2	9
Planar transformer	1 + 20%	92%	1 nF	58 A/mm^2	8

8.1.1.2 Reliability Considerations

- Pay close attention to the insulation between windings and the *dV/dt* of the voltage applied to the transformer. At high switching frequencies, ensuring adequate isolation becomes more challenging due to increased *dV/dt* stresses.

8.1.1.3 Application Suitability

- Planar transformers are particularly advantageous when they can be integrated directly into a printed circuit board (PCB), which is common in power supplies with power levels below approximately 500 W.

8.1.2 Conclusions and Recommendations for Power Transformer

The proximity effect is a significant factor influencing conductive losses, as illustrated in Fig. 8.8. To reduce losses caused by the proximity effect, it is advisable to minimize the number of winding layers. This involves keeping the turn ratio as low as possible and reducing the number of turns to the minimum necessary. Attention should also be paid to the waveform of the current and the duty cycle.

Regardless of how complex a power transformer and the low-voltage side of a power converter may be, a design with a single transformer and low-voltage side typically results in a smaller, more cost-effective solution with better reliability compared to configurations with multiple transformers or parallel power stages.

For applications involving three or four ports in an isolated DC–DC converter, it is generally more effective to design a separate transformer for each port rather than using a single power transformer with multiple windings. This approach simplifies the design and reduces costs. In summary, a cost-effective power transformer design should be as simple as possible.

8.2 Power Inductor

The power inductor is one more crucial component in DC–DC converters. Inductors should be categorized according to the specific part of the converter they serve.

8.2.1 High-Voltage Side Considerations

On the high-voltage side of a DC–DC converter, differential and common mode inductors are often readily available for purchase. These components typically handle lower currents and can be relatively straightforward to design.

Another type of inductor used on the high-voltage side is the AC inductor, commonly employed as a resonant or commutation inductor in topologies like phase-shifted converters. Litz wire is an excellent choice for winding these inductors due to its ability to reduce skin and proximity effects, while ferrite is a suitable material for the core. The core should have low permeability, which may require the use of air gaps or distributed gaps. It is important to note that all AC magnetic components generate electromagnetic fields, which can cause interference. This is especially true for inductors, necessitating careful consideration of their placement and, if needed, the design of special shielding to mitigate interference, as illustrated in Fig. 8.19. However, such measures can increase the overall cost.

The power transformer inherently has leakage inductance, which should be utilized as much as possible to avoid the need for additional inductors. When designing commutation inductors, it's crucial to consider the specific topology being used. In some cases, it may be possible to modify the topology to either eliminate the need for a commutation inductor or reduce the inductor's specifications. This approach can optimize performance and potentially reduce costs.

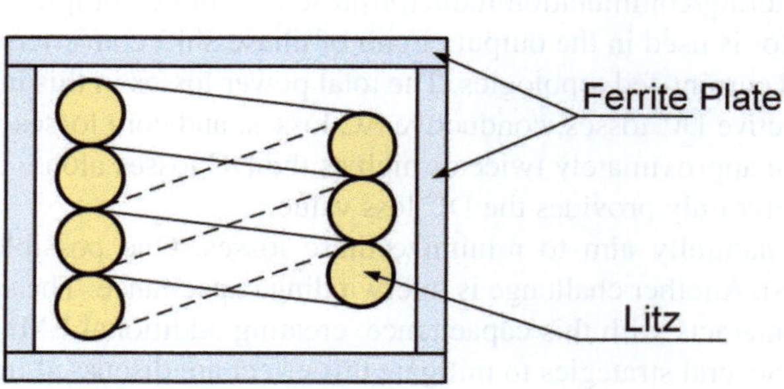

Fig. 8.19 AC inductor enclosed in shield created with ferrite plates

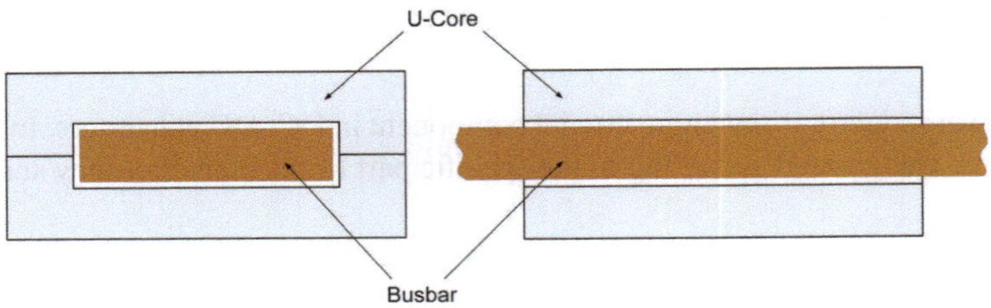

Fig. 8.20 Differential inductor—busbar and U-cores

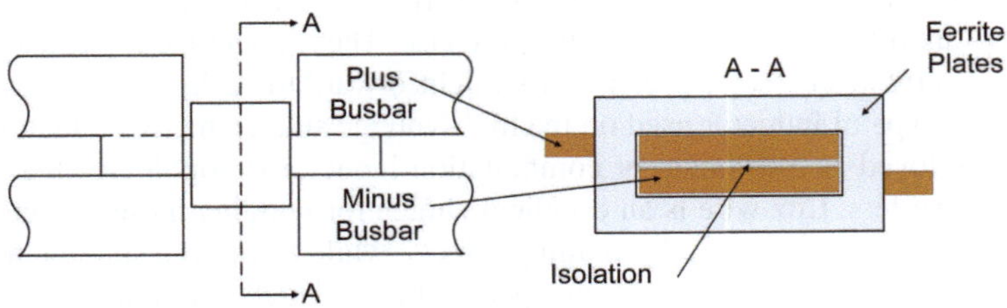

Fig. 8.21 One method for implementing a common mode inductor

8.2.2 Low-Voltage Side Considerations

On the low-voltage side, the picture differs significantly due to the high current requirements. Consequently, all types of inductors necessitate special mounting and high-current connections. In this area, we have three types of inductors: differential, common mode, and filtering/commutation inductors.

Differential inductors can be implemented relatively simply using a clamp/U-core, such as the 00K4119U090, as shown in Fig. 8.20. In contrast, common mode inductors are more complex to implement because they require the busbars (both positive and negative) to be combined. Figure 8.21 illustrates one method for implementing a common mode inductor. Ferrite plates can be used around the busbars to aid in the design. This design approach is not lightweight, which necessitates the use of support stands. This requirement significantly impacts the cost of the converter. The filtering/commutation inductor presents a more complex scenario. This type of inductor is used in the output circuit of phase-shift converters and the input circuit of dual current-fed topologies. The total power losses in this inductor are the sum of conductive DC losses, conductive AC losses, and core losses. Typically, the total losses are approximately twice as high as the DC losses alone. Unfortunately, the manufacturer only provides the DC loss value.

Engineers naturally aim to minimize these losses. One possible approach is detailed in [18]. Another challenge is interwinding capacitance. The signal from the power stage interacts with this capacitance, creating additional EMI in the output/input circuit. Several strategies to mitigate this effect are discussed in [8, 19, 20].

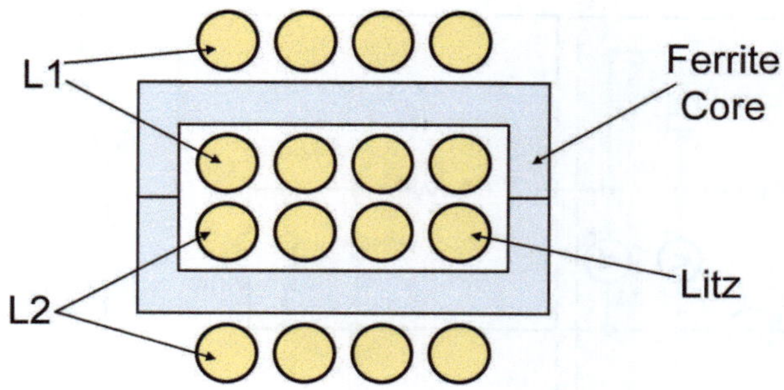

Fig. 8.22 One possible implementation of the autotransformer for the dual current-fed topology

One effective method to reduce capacitance is to use split inductors. This involves using two inductors, each with half the value of the original inductor, connected in series. This configuration results in a capacitance that is at least twice as small and allows for better cooling due to the increased surface area. This concept is illustrated in Chap. 6 for the phase-shift topology. Generally, this design helps reduce costs primarily by increasing the quantity of inductors purchased. Another engineering solution is the use of autotransformers, as demonstrated by the authors for dual current-fed and dual current-fed with switch capacitor applications. Figure 8.22 illustrates one possible implementation of this autotransformer for the dual current-fed topology, which is discussed in Chap. 6.

This autotransformer requires a U-core with a small air gap, approximately 0.1 mm. The winding uses LITZ wire, 8 AWG, with two wires in parallel, each winding placed on a different leg of the core. The losses with the autotransformer are approximately half of those with traditional inductors, and its capacitance is around two to three times smaller. Additionally, the volume is reduced by a factor of 4–5. Regarding cost, it is a matter of comparison. The autotransformer generally costs less than a custom inductor. However, when compared to off-the-shelf inductors, the cost-effectiveness may vary.

8.3 Measuring Magnetic Components

Magnetic components, such as power transformers and inductors, are custom-designed and require specific connections to the power stage in high-power DC–DC converters, especially when the current through these components exceeds hundreds of amps. This necessitates thorough testing and validation during the development process. If the results are unsatisfactory, there is an opportunity to analyze, modify the design, and retest, which ultimately saves time in the converter development cycle. This need for testing is well understood by design engineers, and there are numerous publications that detail the methods for measuring conductive and core losses. The author has selected several relevant references [21–25] that provide

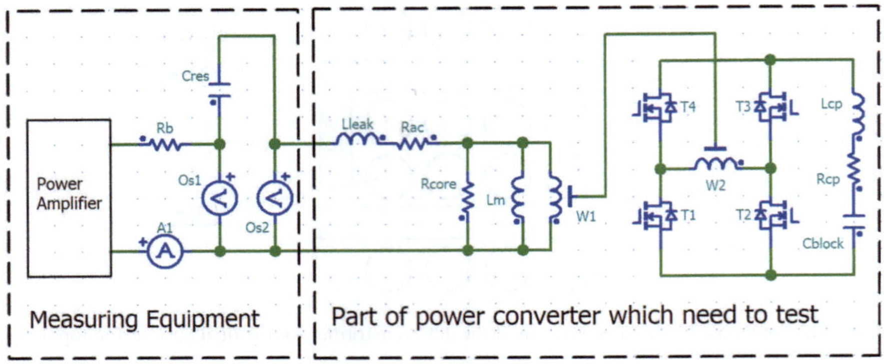

Fig. 8.23 Schematic of a test setup that uses a resonance method to test high-frequency transformers, including low-voltage side of power stage

valuable insights into these measurement techniques. While the references are more suited for research purposes than for practical applications, they provide valuable knowledge.

The author utilized and recommended the resonance method for testing, which involves measuring the power transformer together with the low-voltage power stage. Figure 8.23 shows the schematic setup of the resonance method, including the transformer that needs to be tested and the low-voltage side of the power stage.

During testing, a DC voltage should be applied to the gates of transistors T1 and T3 or T2 and T4, causing the transistors to behave like resistors with a resistance value equal to their Rdc-ON. All measurements should be conducted at a frequency equal to the commutation frequency. An LCR meter is sufficient for preliminary measurements and estimating the parameters of the power transformer and low-voltage power stage. The components of the power converter that need to be tested include:

- **Lleak**: the leakage inductance of the transformer
- **Rac**: the winding resistance of the transformer
- **Lm**: the magnetizing inductance
- **Rcore**: representing the core losses

Additionally, **Rcp**, **Lcp**, and **Cblock** represent the resistance, inductance, and capacitance of the low-voltage side components, including connections, PCB traces, and output capacitors.

The measuring equipment included a power amplifier capable of providing a minimum of 4 A current to a 2-ohm load and operating at a sinewave frequency of up to 400 kHz. For this purpose, the author used an audio power amplifier, specifically a Hafler model, with two channels connected in parallel. This amplifier uses MOSFETs in its output stage. Other amplifiers can be used, provided they offer similar performance. The resistor **Rb** is a non-inductive power resistor with a value of 2–4 Ω. The capacitor **Cres** is the resonance capacitor, whose value is adjusted to match the test frequency, which is equal to the commutation frequency intended for the transformer in the converter. This capacitor should have a minimum ESR (Equivalent Series Resistance) for optimal performance.

Os1, **Os2**, and **A1** (current transformer connected to the oscilloscope) are channels on the oscilloscope. When the test is running in resonance, dividing the voltage measured by Os1 by the current measured by A1 provides the resistance of the power converter components under test, including the ESR of the resonance capacitor. The product of A1 and Os1 gives the power consumed by the power converter components being tested. This measurement allows for the calculation of power losses in the power transformer as well as the power losses on the low-voltage side.

The current on the low-voltage side is determined by multiplying the current measured by A1 by the transformer ratio. For instance, if the current from the power amplifier is 10 A and the transformer ratio is 10, the current on the low-voltage side would be 100 A. This method allows design engineers to easily measure each component on the low-voltage side and identify areas with unacceptable levels of power loss. This process helps to reduce the development time of the power converter and achieve optimal performance.

The same process can be used to measure core losses and AC losses in the inductor. When the low-voltage side is an open circuit (i.e., no DC voltage is applied to the gates of the transistors), the measured power will primarily indicate core losses plus conductive losses (Rac). This measurement provides a practical approximation. Additionally, the voltage measured by Os2 can be used to calculate the flux density, which is ideally set to match the flux density that will be used in the actual converter. In this case, the resonance capacitor is selected to resonate with the magnetizing inductance.

8.4 Conclusion/Note

Power magnetics are integral components of a power converter and can only be effectively utilized within the context of the entire converter system. Therefore, design engineers must consider the overall system design and topology when selecting and designing these components. It is crucial to take into account the interplay between the power magnetics and other parts of the power converter to ensure optimal performance and reliability.

8.5 Supplemental

The content of this chapter represents the original, creative work of the authors. We utilized OpenAI's ChatGPT to check grammar, spelling, and improve the clarity of certain portions of the text. Some of the images and data contained in this chapter were provided by Havis, Inc., with support from Alec Cook. Portions of the text and images are from tutorials that we presented at the *2017 IEEE Energy Conversion Congress and Exposition* (ECCE 2017), Cincinnati, OH, USA, 2017 and *2022 IEEE Energy Conversion Congress and Exposition (ECCE 2022),* Detroit, MI, USA, 2022

References

1. M.K. Kazimierczuk, *High-Frequency Magnetic Components*, 2nd edn. (Wiley, 2014)
2. W.M. Colonel, T. Mclyman, *Transformer and Inductor Design Handbook*, Revised and expanded, 3rd edn. (Marcel Dekker, Inc., 2004)
3. W.G. Hurley, W.H. Wolfle, *Transformers and Inductors for Power Electronics Theory, Design and Applications* (Wiley, 2013)
4. S.C. Bell, P.S. Bodger, Power transformer design using magnetic circuit theory and finite element analysis—a comparison of techniques, in *2007 Australasian Universities Power Engineering Conference*, (2007)
5. E.I. Amoiralis, M.A. Tsilia, P.S. Georgiakis, The state of the art in engineering methods for transformer design and optimization: a survey. J. Optoelectron. Adv. Mater. **10**(5), 1149–1158 (2008)
6. R. Ridley, *Winding Capacitance* (APEC Magnetics Industry Session, 2018)
7. H. Zhang, S. Wang, Y. Li, Q. Wang, D. Fu, Two-capacitor transformer winding capacitance models for common-mode EMI noise analysis in isolated DC–DC converters. IEEE Trans. Power Electron. **32**(11), 8458 (2017)
8. S. Wang, F.C. Lee, Analysis and applications of parasitic capacitance cancellation techniques for EMI suppression. IEEE Trans. Ind. Electron. **57**(9), 3109 (2010)
9. Z. Shen, M. Kaymak, H. Wang, J. Hu, L. Jin, F. Blaabjerg, R.W. De Doncker, The faraday shields loss of transformers. IEEE Trans. Power Electron. **35**(11), 12194 (2020)
10. High-Frequency Power Transformer, Patent US 7,123,123
11. Integral Planar Transformer and Busbar, Patent US 8,237,535 B2
12. Integrated Magnetic Module, Patent US 9,490,057 B2
13. Method of Forming a Transformer Winding, Patent US 10,679,784 B1
14. C. Fei, F.C. Lee, Q. Li, High-efficiency high-power-density 380V/12V DC/DC converter with a novel matrix transformer, in *IEEE Applied Power Electronics Conference and Exposition (APEC)*, (2017)
15. M. Mingkai, F.C. Lee, Design and optimization of a 380–12 V high-frequency, high-current LLC converter with GaN devices and planar matrix transformers. IEEE J. Emerg. Select. Topics Power Electron. **4**(3) (2016)
16. Matrix Planar Transformer, Patent US 10,074,474 B2
17. B.W. Carsten, The low leakage inductance of planar transformers; fact or myth? in *IEEE Applied Power Electronics Conference and Exposition (APEC)*, (2001)
18. M. Nymand, U.K. Madawala, M.A.E. Andersen, B. Carsten, O.S. Seiersen, Reducing AC-winding losses in high-current high-power inductors, in *Annual Conference of IEEE Industrial Electronics Conference Paper*, (2009)
19. Cancelation of Inductor Capacitance, Patent US 7,554,423 B2, 30 Jun 2009
20. S. Wang, F.C. Lee, J.D. van Wyk, Inductor winding capacitance cancellation using mutual capacitance concept for noise reduction application. IEEE Trans. Electromagn. Compat. **48**(2), 311 (2006)
21. M. Mingkai, F.C. Lee, A new high frequency inductor loss measurement method, in *2011 IEEE Energy Conversion Congress and Exposition*, (2011)
22. Y. Han, W. Eberle, Y.-F. Liu, A practical copper loss measurement method for the planar transformer in high-frequency switching converters. IEEE Trans. Ind. Electron. **54**(4), 2276 (2007)

23. Y. Han, Y.-F. Liu, A practical transformer Core loss measurement scheme for high-frequency power converter. IEEE Trans. Ind. Electron. **55**(2), 941 (2008)
24. M. Mingkai, Q. Li, D.J. Gilham, F.C. Lee, K.D.T. Ngo, New core loss measurement method for high-frequency magnetic materials. IEEE Trans. Power Electron. **29**(8), 4374 (2014)
25. D. Hou, M. Mu, F.C. Lee, Q. Li, New high-frequency core loss measurement method with partial cancellation concept. IEEE Trans. Power Electron. **32**(4), 2987 (2017)

Chapter 9
Advanced Concepts

The goal of this book is to explore the practical implementation of isolation DC–DC converters. These converters have a broad range of applications, but there are three areas where their use is particularly significant. This chapter will specifically address the practical implementation and operation of isolation DC–DC converters in these areas:

- **Bidirectional DC–DC converters**: These converters are increasingly being used in a variety of applications.
- **Multiport converters**: Like bidirectional converters, multiport converters are also seeing wider adoption.
- **Multilevel configuration of power stages in DC–DC converters**: This configuration is another important area of focus for the practical application of these converters.

9.1 Bidirectional DC–DC Converter

One application area for isolated DC–DC converters is in enabling bidirectional power conversion. While this book does not focus on the detailed description of bidirectional converters, we highlight their use in applications where power needs to flow in both directions. Typically, a battery is present on one side of the converter, such as in renewable energy systems or distribution systems [1, 2]. In one direction, the converter charges the battery, while in the opposite direction, the battery discharges power to the load. In these applications, it is crucial to ensure that the converter can handle a voltage range on the battery side that is at least 1.5 times the nominal voltage. Additionally, during the charging process, the converter operates in a current source mode. Another scenario occurs when batteries are present on both sides of the converter, a common application in electric and hybrid

A. Isurin, M. J. Scott, *Practical Design Considerations for Isolated DC-DC Converters*, https://doi.org/10.1007/978-3-031-94427-7_9

transportation systems [3]. In this case, the converter operates in current source mode during power conversion in both directions. The converter must be able to handle a voltage range that is at least 1.5 times the nominal voltage on each side. This requirement implies that the converter needs to manage a regulation range of at least 2.25 times the nominal voltage, which can be challenging, particularly when dealing with low voltage and high current on one side. Various other configurations and applications are discussed in [4, 5].

Theoretically, all types of power converters can operate bidirectionally, but a true bidirectional converter must be able to boost the voltage in one or both directions. The challenge is how to implement this practically and cost-effectively. The following approaches can be considered for achieving this requirement. The Dual Active Bridge (DAB) is inherently bidirectional, but it has significant limitations, especially when handling currents over 100 A. It is also not cost-effective. The DAB operates within a regulatory framework where regulation is managed by a single control section—the entire power stage. This design makes the DAB versatile, but like many universal solutions, it comes with numerous disadvantages. In general, increasing universality tends to decrease quality, and improving quality often means reducing universality. Book reference [6] discusses various applications of this topology, but mainly for systems around 1 kW.

We explore other directions and solutions. The first approach is to use regulation sections on both sides of the transformer, following the philosophy of "divide and conquer." This approach is exemplified by the dual current-fed topology, its modifications, and the CLLC topology, which are discussed in references [7–11]. In these configurations, boosting gain is achieved by creating a short circuit on the high-voltage side when the low-voltage side is the primary side. The second approach involves combining two different types of control: variable frequency and PWM control [12, 13]. This method, which is detailed in Chap. 10, illustrates a practical way to enhance converter performance and flexibility.

The third approach involves using a reconfigurable high-voltage low-current side along with various types of control regulation. For instance, configurations can switch from a half-bridge to a full-bridge, from a full-bridge to a half-bridge, or involve reconnecting taps of a power transformer or using multiple transformers [14]. This reconfiguration is controlled using a data table in a feed-forward mode with a hysteresis loop. It is also possible to combine any of the approaches mentioned above [15–17].

All the approaches discussed above illustrate that bidirectional power converters are specialized and do not have a straightforward implementation path. In other words, the design and implementation of these converters are highly individualized, depending on the specific technical requirements that need to be met. Consequently, there are numerous variations in how bidirectional power converters can be realized.

9.2 Multiport Operation

Another type of DC–DC converter application is in multiport converters. These converters are used in various fields, including transportation [18–20], renewable energy systems [21–23], microgrids [24, 25], and other areas [26, 27]. The use of multiport converters is increasing due to their versatility and efficiency in managing multiple power sources or loads. Generally, multiport converters can be classified into three different types based on their implementation. The first type involves non-isolated ports, where there is no galvanic isolation between the ports. The second type uses complete galvanic isolation, typically achieved with transformers. There is also an intermediate form of isolation using capacitors, which the author considers non-isolated since capacitors do not block AC current. The third type includes any combination of the first two methods, providing flexibility in design based on specific requirements.

The target of this book is to explore practical and high-current isolated DC–DC converters. Therefore, we will examine the differences between isolated and non-isolated DC–DC converters. Non-isolated converters have several significant disadvantages compared to transformer-isolated converters in terms of electrical behavior.

Figure 9.1 illustrates the main disadvantage of a specific type of non-isolated converter: the issue of parasitic transient currents. This problem becomes significant when the converter operates with switching currents above 20–30 A. At currents of 100 A and beyond, this issue requires extremely careful attention. This transient current generates high levels of EMI, necessitating the use of more complex EMI filters on both the input and output sides of the converter. The consequence of this "treatment" is reduced efficiency, increased size and weight, and higher costs. Another approach to mitigating these issues is the use of a multiphase topology, as shown in Fig. 9.2.

However, this method introduces challenges in current sharing and loop control, and it can be quite expensive due to the need for parallel connections of three or more power stages.

Another solution is to use an isolated DC–DC converter. For converters with currents exceeding 100 A, the magnetic components are typically custom-designed. When comparing filtering/commutation inductors with power transformers in terms

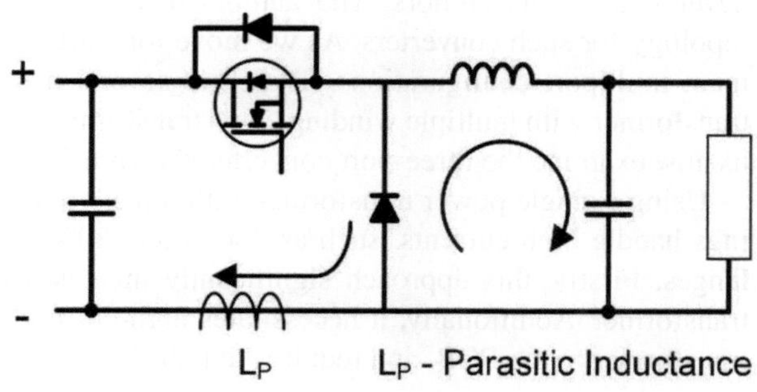

Fig. 9.1 Schematic of a buck converter where the stray inductance (L_p) produces a parasitic transient current (PTC)

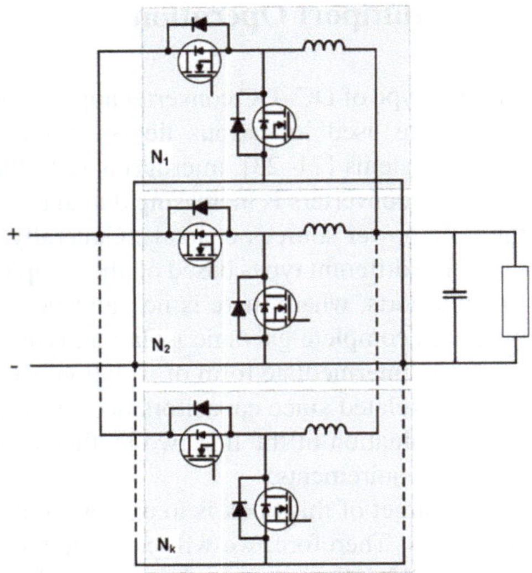

Fig. 9.2 Block diagram of a method to solve the PTC problem using parallel power stages

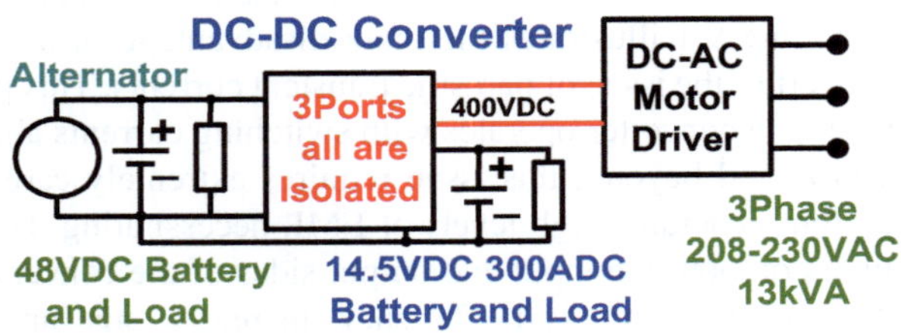

Fig. 9.3 The three-port converter for vehicle electrification

of weight, dimensions, and overall efficiency, transformers often prove to be superior. In Chap. 8, the authors emphasize the importance of avoiding the use of filtering/commutation inductors whenever possible, due to considerations of efficiency, weight, and size—parameters that are particularly critical in the aviation industry. Finally, isolated DC–DC converters can offer cost advantages over non-isolated converters, especially when a high ratio between input and output voltage is required.

Figure 9.3 illustrates a three-port converter designed for vehicle electrification, developed by the authors. The authors recommend using an isolated multiport topology for such converters. As we move forward, we will explore how to implement multiport configurations. There are several possible approaches: using one transformer with multiple windings, two transformers, or multiple transformers. Let us now examine the three-port converter shown in Fig. 9.3.

Using a single power transformer with three windings—where two of the windings handle high currents, such as 300 A and 300–600 A—presents several challenges. Firstly, this approach significantly increases the size and volume of the transformer. Additionally, it necessitates high-current connections from the power transformer to the PCB, and requires that the high-current power section be located close to the transformer. This setup can complicate the design and integration, making it less practical for many applications. The second consequence of using a single power transformer with high-current windings is reduced efficiency, primarily due to excessively long traces on the PCB. Additionally, this approach significantly increases both the cost of the converter and the labor costs on the production floor, as the unit becomes more complex to manufacture. Therefore, the author decided to use two transformers [27].

The conclusion is that for configurations involving one or two transformers, it is advisable to avoid high-power transformers—particularly those over several kVA—with multiple windings and high-current windings, especially when implemented with foil.

9.3 Multilevel Topology

What is multilevel topology? In essence, it refers to a configuration where active switches are connected in series. Another definition describes a power leg as having four or more active switches. For the purposes of this book, which focuses on practical considerations for isolated DC–DC converters, we will adhere to the following definitions and assumptions. Design engineers must consider several key requirements when designing electronic equipment, especially for the automotive industry. These include cost, reliability, and electromagnetic compatibility (EMI). The primary requirements for electronic equipment in the automotive industry are cost, reliability, and electromagnetic compatibility (EMI). A simple illustration of these requirements is that the half-bridge topology is generally preferred over the full-bridge topology under similar conditions. This preference is because a converter with two active switches typically has lower overall cost compared to a converter with four active switches. Additionally, a converter with two active switches is generally more reliable than one with four active switches.

The practical consequence of these rules is that, under the same conditions, the LLC topology with a half-bridge configuration is generally superior to the phase-shift topology. Additionally, the LLC topology typically results in lower electromagnetic interference (EMI) compared to the phase-shift topology. The LLC topology offers soft switching across the entire output power range and on both sides of the power stage, including both low- and high-voltage sides. Some sources [28] suggest that using cheaper devices connected in series can reduce costs and potentially improve switching performance. However, this claim is questionable, as the comparison in [28] is between half-bridge and full-bridge topologies. There are also many other methods available to enhance switching performance.

We are comparing the cost of one power leg with different numbers of active switches connected in series. Table 9.1 presents this comparison. The table includes three configurations: the first leg has two active switches, the second leg has four active switches connected in series, and the third leg has eight active switches connected in series.

Table 9.1 Cost comparison for the main components for one power leg with different numbers of active switches connected in series

MOSFETs	Rdc-ON each	Gate driver and PS $3.5 driver and PS	Total cost
IMYH200R050M1HXKSA1 34 × 2 = 68 (cost-250p)	50 mΩ	2 × 3.5 = 7	$75.0
IMW120R020M1HXKSA1 19.56 × 4 = 78.24 (cost-500p)	27 mΩ	4 × 3.5 = 14	$92.24
AIMZA75R016M1HXKSA1 16.45 × 8 = 131.6 (cost-500p)	15 mΩ	8 × 3.5 = 28	$159.6

This comparison is based on the following assumptions: the DC bus voltage is the same for all legs, and all legs have similar conductive losses. Switching losses and switching times are not considered, as only soft switching is used. The cost does not include the additional expenses associated with increasing the quantity of components on the PCB. In other words, real implementations will experience a significantly higher cost and a notable reduction in reliability. Another implementation of multilevel topologies, where multiple active switches are connected in series, is used to achieve high voltage gain [29]. In such cases, the power transformer is generally more cost-effective and reliable. Additionally, the authors strongly believe that cost-effective isolated DC–DC converters with output power below 1 kW can be effectively implemented using a maximum of two active switches, excluding synchronous rectifiers from the consideration.

9.4 Conclusion

Design engineers must have very strong justifications for using multilevel configurations, especially when the DC bus voltage is less than 1000 VDC and when wide-bandgap (WBG) transistors are available. The discussion above illustrates that engineers need to be cost-conscious and skilled in managing budgets effectively.

9.5 Supplemental

Some of the images and data contained in this chapter were provided by Havis, Inc., with support from Alec Cook.

Portions of the text and images are from tutorials presented at the *2017 IEEE Energy Conversion Congress and Exposition* (ECCE 2017), Cincinnati, OH, USA, 2017 and *2022 IEEE Energy Conversion Congress and Exposition (ECCE 2022)*, Detroit, MI, USA, 2022.

The content of this chapter represents the original, creative work of the authors. We utilized OpenAI's ChatGPT to check grammar and spelling and improve the clarity of certain portions of the text.

References

1. L. Zheng, X. Yan, Y. Fang, L. Sun, Mode analysis and optimum design of bidirectional CLLC resonant converter for high-frequency isolation of DC distribution systems, in *IEEE Energy Conversion Congress and Exposition (ECCE), Conference Paper*, (2015)
2. G. Li, J. Huang, W. Zhang, X. Tong, J. Xiao, M.K.L. Hai, High-efficiency oriented control method of bidirectional CLLC resonant DC/DC converter for energy storage system, in *15th IEEE Conference on Industrial Electronics and Applications (ICIEA), Conference Paper*, (2020)
3. E. Martinez-Vera, P. Banuelos-Sanchez, Review of bidirectional DC-DC converters and trends in control techniques for applications in electric vehicles. IEEE Lat. Am. Trans. **22**(2), 144 (2024)
4. S.A. Gorji, H.G. Sahebi, M. Ektesabi, A.B. Rad, Topologies and control schemes of bidirectional DC-DC power converters: an overview. IEEE Access **7** (2019)
5. N.M.L. Tan, T. Abe, H. Akagi, Topology and application of bidirectional isolated DC-DC converters, in *8th International Conference on Power Electronics—ECCE Asia*, (2011)
6. Deshang Sha, Guo Xu, (2019). High-Frequency Isolated Bidirectional Dual Active Bridge DC–DC Converters with Wide Voltage Gain
7. X. Pan, H. Li, Y. Liu, T. Zhao, J. Chenchen, A.K. Rathore, An overview and comprehensive comparative evaluation of current-fed-isolated-bidirectional DC/DC converter. IEEE Trans. Power Electron. **35**(3), 2737 (2020)
8. D. Sha, X. Yaxiong, Y. Jiankun Zhang, Yan., Current-fed hybrid dual active bridge DC–DC converter for a fuel cell power conditioning system with reduced input current ripple. IEEE Trans. Ind. Electron. **64**(8), 6628 (2017)
9. X. Li, J. Huang, Y. Ma, X. Wang, J. Yang, W. Xiaojie, Unified modeling, analysis, and design of isolated bidirectional CLLC resonant DC–DC converters. IEEE J. Emerg. Select. Topics Power Electron. **10**(2), 2305 (2022)
10. Bi-Directional Isolated DC-DC converter for Electrification of Transportation, Patent US 10,483,862 B2, 19 Nov 2019
11. Bi-Directional Isolated DC-DC converter for Electrification of Transportation, Patent US 11,063,518, 13 July 2021
12. A.M. Pernía, M.J. Prieto, P.J. Villegas, F. Nuño, J.A. Martín-Ramos, LCLC resonant converter with PWM control, in *21st European Conference on Power Electronics and Applications (EPE '19 ECCE Europe), Conference Paper*, (2019)
13. W. Hua, H. Wu, Z. Yu, Y. Xing, K. Sun, A phase-shift modulation strategy for a bidirectional CLLC resonant converter, in *10th International Conference on Power Electronics and ECCE Asia (ICPE 2019—ECCE Asia), Conference Paper*, (2019)
14. X. Yang Xuan, W.C. Yang, T. Liu, X. Hao, A novel three-level CLLC resonant DC–DC converter for bidirectional EV charger in DC microgrid. IEEE Trans. Ind. Electron. **68**(3), 2334 (2021)
15. Y. Shen, H. Wang, A. Al-Durra, Z. Qin, F. Blaabjerg, A bidirectional resonant DC–DC converter suitable for wide voltage gain range. IEEE Trans. Power Electron. **33**(4), 2957 (2018)
16. J.-W. Lim, M. Kim, Bidirectional soft switching push–pull resonant converter over wide range of battery voltages. IEEE Trans. Power Electron. **36**(11) (2021)
17. C. Bai, M. Kim, Bidirectional resonant converter with minimized switching loss over wide operating voltage range. IEEE J. Emerg. Select. Topics Power Electron. (2022)

18. K. Itoh, S. Inoue, M. Ishigarki, T. Sugiyama, T. Umeno, Analysis and Design of Multiport DC-DC converter for next-generation hybrid-vehicle subsystems. R&D Review of Toyota CRDL **48**(2) (2017)
19. S. Chakraborty, V. Hai-Nam, M.M. Hasan, D.-D. Tran, M. El Baghdadi, O. Hegazy, DC-DC converter topologies for electric vehicles, plug-in hybrid electric vehicles and fast charging stations: state of the art and future trends. Energies **12**(8), 1569 (2019)
20. A.M. Fares, C. Klumpner, M. Sumner, A novel multiport DC-DC converter for enhancing the design and performance of battery–supercapacitor hybrid energy storage systems for unmanned aerial vehicles. Applied Sciences **12**(6), 2767 (2022)
21. R. Sekar, D.S. Suresh, H. Naganagouda, A review on power electronic converters suitable for renewable energy sources, in *International Conference on Electrical, Electronics, Communication, Computer, and Optimization Techniques (ICEECCOT), Conference Paper*, (2017)
22. R.M. Al-Dalawi, A. Al-Dujaili, D.A. Pereira, A comprehensive review of multi-port DC/DC converters for the off-grid system integration with renewable energy resources. J. Tech. **5**(2) (2023)
23. E. Babaei, O. Abbasi, S. Sakhavati, An overview of different topologies of multi-port DC/DC converters for DC renewable energy source applications, in *13th International Conference on Electrical Engineering/Electronics, Computer, Telecommunications and Information Technology (ECTI-CON), Conference Paper*, (2016)
24. A.A. Saafan, V. Khadkikar, M.S. El Moursi, H.H. Zeineldin, A new multiport DC-DC converter for DC microgrid applications, in *IEEE Industry Applications Society Annual Meeting (IAS), Conference Paper*, (2021)
25. P. Kolahian, H. Tarzamni, A. Nikafrooz, M. Hamzeh, Multi-port DC–DC converter for bipolar medium voltage DC micro-grid applications. IET Power Electron. **12**, 1841 (2019)
26. A.H. Chander, L.K. Sahu, P.T. Bankupalli, Review on general architecture and selection of multiple input converters, in *2020 First International Conference on Power, Control and Computing Technologies (ICPC2T), Conference Paper*, (2020)
27. A.M. Elmakawi, K.Ç. Bayındır, Non-isolated multi-port inverter topologies for renewable energy applications: a review, in *1st Global Power, Energy and Communication Conference (GPECOM), Conference Paper*, (2019)
28. J. Khodabakhsh, G. Moschopoulos, A study of multilevel resonant DC-DC converters for conventional DC voltage bus applications, in *Applied Power Electronics Conference and Exposition (APEC), Conference Paper*, (2018)
29. A.K. Singh, A.K. Mishra, K.K. Gupta, Y.P. Siwakoti, High voltage gain bidirectional DC-DC converters for supercapacitor assisted electric vehicles: a review. CPSS Trans. Power Electron. Appl. **7**(4), 386 (2022)

Chapter 10
Practical Direction of Developing New Converter

In this chapter, the authors illustrate practical engineering methods for developing a new isolated bidirectional DC–DC converter. We will demonstrate how the recommendations from previous chapters are applied while adhering to key engineering principles: cost-effectiveness, reliability, and compatibility. Of course, the authors show their process of developing the new converter, but this does not necessarily mean that it is the best approach. Other engineers may take different paths, but the main principles of cost, reliability, and compatibility must always be applied.

10.1 Technical Specifications of the Converter to Be Developed

The application is for renewable energy, with the low-voltage side connected to a lithium-ion battery.

- **Bidirectional operation**
- **Low-voltage side**: Battery, 40–60 VDC
- **High-voltage side**: 380–420 VDC
- **Step-up mode (discharging the battery)**: Output voltage of 380–400 V. Maximum input current of 80 A, with an output power of up to 4 kW
- **Step-down mode (charging the battery)**: Input voltage of 400–420 V. Maximum output/charge current of 40 A
- **Operating conditions**: No active cooling system, relying on natural convection. Ambient temperature range of 0–50 °C
- **Cost considerations**: The converter must be cost-effective
- **Compliance**: CISPR 25 Class 4
- **Lifetime**: Minimum of 10 years

A. Isurin, M. J. Scott, *Practical Design Considerations for Isolated DC-DC Converters*, https://doi.org/10.1007/978-3-031-94427-7_10

10.2 Design Considerations

The first step in the development is that the converter must use soft-switching technology. The requirement for cooling by natural convection demands achieving the highest possible efficiency, which helps minimize the size and cost of the heatsink. For a cost-effective design, a reasonably high level of commutation frequency should be used, which helps reduce the size and cost of filters and the power transformer. For this power level, the switching frequency should be between 250 and 500 kHz. This frequency range also facilitates easier compliance with CISPR 25 Class 4 standards. Now that we have determined the main operating conditions, we can move on to the next critical step: selecting the topology to use or developing something new. The dual current-fed topology with active clamping is not suitable due to its maximum switching frequency limit of 200 kHz. Our target is a switching frequency of around 300 kHz or higher. While the dual current-fed topology with a switched capacitor is ideal for this application, it cannot be used due to an existing patent owned by another company.

10.2.1 Specific Requirements for This Converter Application

1. **Step-up mode**:
   - Operates as a voltage source with a current limit, and the output voltage can vary between 380 and 400 V.
2. **Step-down mode**:
   - Functions as a current source with a maximum current of 40 A and an input voltage range of 400–420 V.

10.3 Evaluation of Topologies

When addressing these tasks, cost-effectiveness is the primary consideration. Currently, two topologies are available: CLLC and Resonant with PWM control. Neither of these options is perfect. The CLLC topology and its variations perform well as long as the voltage range does not exceed 1.3 times its nominal value. However, when the voltage range reaches 1.5 times (e.g., 60 V/40 V = 1.5) or even twice the nominal value, efficiency decreases significantly [1–3], which ultimately raises the total cost of the converter. In addition, when we talk about the reliability of the converter in terms of parameters, rather than through control or compensation, it is important to maintain optimal energy transfer performance to ensure adequate protection. In other words, reliability should be based on system parameters rather than control adjustments.

The bidirectional Resonant with PWM converter [4, 5] is not suitable for power levels over 3 kW due to its high peak current, which limits the maximum commutation frequency and increases the stress on semiconductors. When the commutation frequency exceeds 100 kHz, efficiency starts to decrease. All these factors negatively impact the total cost of the converter.

Given these considerations, the authors had to choose between developing a completely new topology or using a combination of CLLC and Resonant PWM. Developing a new topology always carries risks and requires careful evaluation due to its high cost and extended development time. Therefore, the preference was to use existing technology wherever possible.

A decision was made to use a combination of CLLC and Resonant PWM. Figure 10.1 shows the schematic of this topology with a full bridge on the low-voltage energy storage side, while Fig. 10.2 shows the schematic with a half-bridge on the low-voltage energy storage side.

The next question is which configuration to use: full bridge or half bridge. Generally, the half-bridge configuration is more cost-effective, but this advantage holds only when the current levels are manageable and the voltage across the power transformer windings is not too low. In the worst-case scenario, the voltage is around 20 V, and the current reaches approximately 200 A, which reduces the benefits of using the half-bridge configuration for this application.

At this stage, we already had an acceptable and ready bridge power stage from a previous project, so the decision was made to use the full-bridge configuration for the prototype to save time and move forward with other aspects of the new design. This choice highlights that engineering often requires compromise to balance cost, time, and performance.

The new proposed topology, shown in Figs. 10.1 and 10.2, overcomes all the disadvantages of the previously described topologies while retaining their advantages. In other words, this proposed topology is a combination of the CLLC Converter and Resonant with the PWM Converter. It features soft-switching across the entire voltage and power range, with a maximum efficiency variation of only 0.4% over the full input/output voltage range.

The design eliminates high current stress, which allows for the use of WBG transistors, enabling commutation frequencies of up to 600–700 kHz. The proposed topology also requires only four active switches in the regulated section, compared

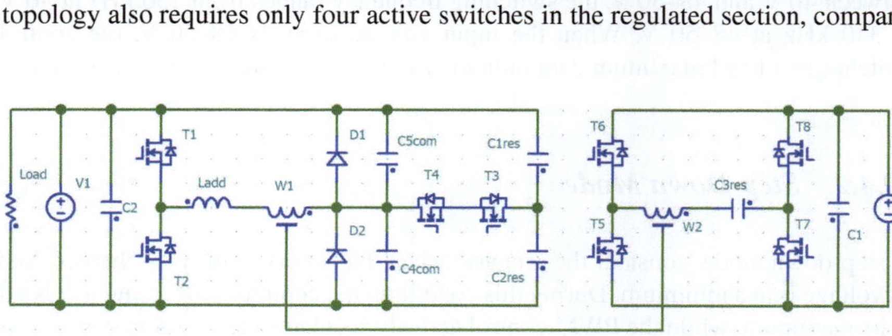

Fig. 10.1 The schematic of the proposed topology with full bridge on energy storage side

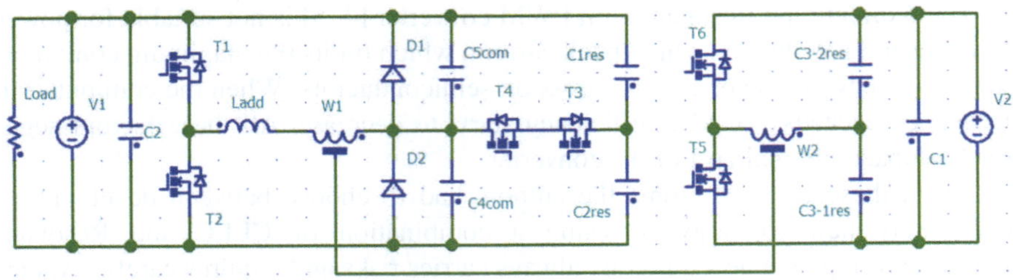

Fig. 10.2 The schematic of proposal topology with half bridge on energy storage side

to the six switches typically needed in the Resonant with PWM Converter [4]. This reduction in components contributes to a more cost-effective design. Let's now look at the operational algorithm of this topology.

10.4 Operation of the Proposed Topology

The proposed topology features two regulation sections: the low-voltage side, which operates as an LLC circuit, and the high-voltage side, which functions as a resonant circuit with PWM control. Together, these sections work as a unified resonant system. Let's briefly review how this topology operates in both step-up and step-down modes.

10.4.1 Step-Up Mode

In step-up mode, the control system operates with a variable switching frequency, depending on the input voltage range, where Vmid/Vmin = 1.2–1.25. During this mode, the PWM section continuously conducts, but it only activates when a rapid reduction of output voltage or current is necessary. In our case, for input voltages between 40 V and 48–50 V, the switching frequency ranges from 250 kHz at 40 V to 340 kHz at 48–50 V. When the input voltage exceeds 48–50 V, the control switches to a fixed maximum commutation frequency, ensuring stable operation.

10.4.2 Step-Down Mode

In step-down mode, consider the scenario when the battery is fully discharged, and its voltage is at a minimum. During this condition, the commutation frequency is set to its maximum, while the PWM control actively regulates the charging current. As the battery charges and its voltage increases, reaching the midpoint value Vmid, the

PWM section remains fully conductive. As the battery voltage continues to rise towards its maximum value, the control system gradually decreases the switching frequency.

In this mode, when the battery voltage is between 40 V and 48–50 V, the commutation frequency is set at 340 kHz. When the battery voltage reaches 60 V, the commutation frequency reduces to 250 kHz.

10.4.3 Important Considerations for Step-Down Mode

In charge mode, when the battery is charged to less than 90% of its capacity, the converter operates in current-source mode. This is highly beneficial as it allows the battery to charge at a constant current level, which extends its lifespan. As the battery reaches above 90% capacity, or close to 100%, the converter should switch to a light-current mode to avoid overcharging. This ensures the battery is fully charged in a controlled manner, which is crucial for the long-term reliability of Li-ion batteries commonly used in renewable energy applications. Overcharging can be dangerous for Li-ion batteries, as they do not tolerate any overvoltage conditions.

Another key safety feature is that the converter must check the battery voltage before initiating the charging process. If the battery voltage is below the minimum acceptable level, the converter will not start charging to prevent potential damage, such as a short circuit in one or more cells.

10.4.4 Conclusion

As we can see, the converter predominantly operates either as a CLLC or as a resonant circuit with PWM control. Detailed descriptions of these modes of operation can be found in references [1–3, 6–9], and we will not repeat them here.

10.4.5 Special Operation Mode of the Converter

The converter can enter a special operation mode when the commutation frequency is not at its maximum and the PWM control is active. This mode functions as a protection mechanism, triggered by rapid and significant changes in load levels or overload conditions in step-up mode. In this operation mode, the topology operates by combining CLLC and PWM modes. In other words, the converter remains in CLLC mode while simultaneously utilizing PWM control. Let's examine this mode in detail.

10.4.6 Operation in Discharge/Step-Up Mode

Figure 10.3 shows the equivalent schematic of the topology in discharge/step-up mode. Here, Lleak represents the leakage inductance of the power transformer, while Ladd is an additional inductor that is part of the resonant inductor. The commutation frequency during this operation is set at 294 kHz. The commutation capacitors *C6com* – *C9com* do not significantly influence the main behavior of the converter; however, they do impact the commutation of transistors T5–T8.

Figure 10.4 displays the simulation waveform voltage and current of the converter when the commutation frequency is within the CLLC mode and regulation is achieved via PWM.

Figure 10.5a–f illustrate the equivalent circuits of each operational time segment. Starting with Fig. 10.5a, which covers the time segment t0–t1: Transistors T5 and T8 turn on at t0, initiating conduction with the current beginning at the magnetizing current value and increasing sinusoidally. This behavior is dictated by the resonant circuit composed of C4com–C5com, Ladd and Lleak, as well as the load value. Simultaneously, the resonance capacitor C3res begins discharging into the primary winding of the transformer. At time t0, devices T5 and T8 start to conduct the magnetizing current in ZVS operation. The body diode of device T2 starts conducting under ZCS during this time. T2, C4com, and C5com remain active during the t0–t1 period.

At t1, the voltages across C4com and C2res equalize, causing T3 and T4 to begin conducting (Fig. 10.5b) under ZVS. Capacitors C1res and C2res become active at t1, with current continuing to rise sinusoidally through T2, T3, T4, and T5, T8. At t2, device T4 turns off, interrupting the current flow while T3 prepares for the next half-cycle. This interruption occurs under ZVS due to T3 and T4 being positioned between C1res, C2res and C3com, C4com (Fig. 10.5c). The commutation voltage and time depend on the values of the commutation capacitors and the current. Higher currents result in lower voltages, thereby reducing EMI levels during the t2–t3 timing.

Between t3 and t4, the inductors Ladd and Lleak begin discharging energy, with current flowing through D1, Ladd, Tr, and T2 (Fig. 10.5d) on the high-voltage side of the converter. Meanwhile, current flows through T5, T6, Tr, Lleak, C3res, and Lmag. On the high-voltage side, the current linearly decreases to zero at t4 as power conversion to the load concludes. On the low-voltage side, the magnetizing current flows through T5, T8, Tr, Lleak, and C3res between t4 and t5 (Fig. 10.5e).

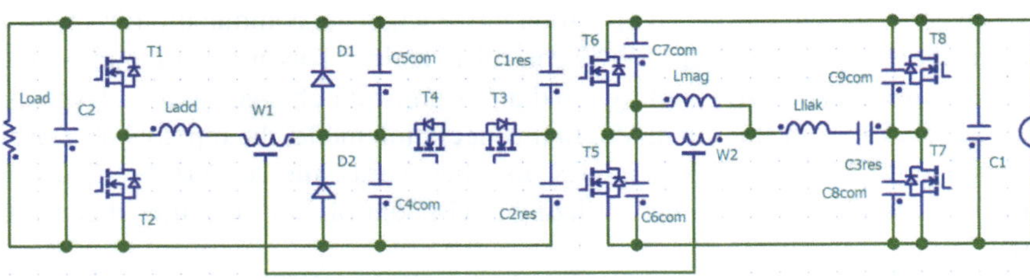

Fig. 10.3 The schematic of proposed topology operating in discharge mode

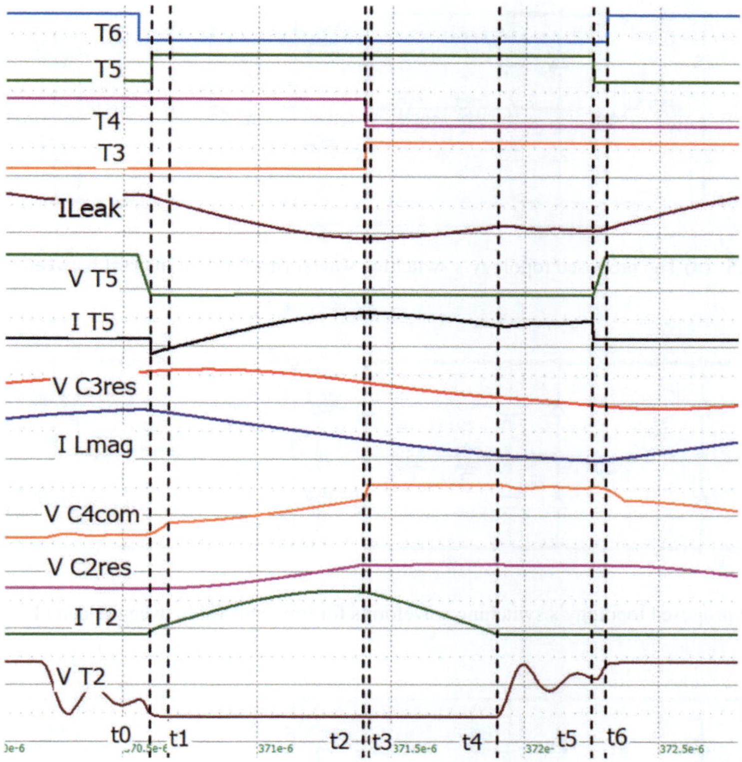

Fig. 10.4 The simulation waveforms of the converter in combination mode CLLC and PWM

During this interval, from t4 to t5, capacitor C3res charges with magnetizing current, preparing for the next half-cycle of power conversion. From t5 to t6 (Fig. 10.5f), T5 and T8 turn off at t5, ceasing the magnetizing current. Capacitors C6com, C7com, C8com, and C9com ensure zero-voltage turn-off for transistors T5 and T8 while facilitating the turn-on of T6 and T7 under zero-voltage conditions. At t6, T6 and T7 switch on, initiating a new power conversion half-cycle that mirrors the previous half-cycle from t0 to t5.

10.5 Calculation of Main Power Components

Now that we understand the principal operations of the new topology/converter, let's consider the preliminary practical calculations for the main power components of this topology, including the resonance tank and the turn ratio of the power transformer. The maximum power of this converter is in step-up mode, so we will calculate all the main components in this mode. This initial engineering calculation will help us estimate the compatibility of this topology.

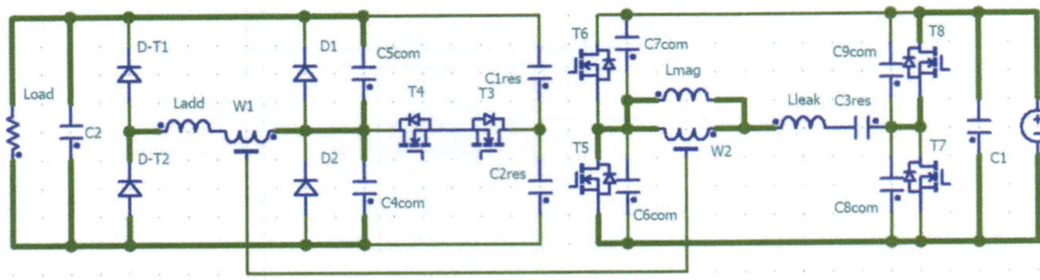

Fig. 10.5 (**a**) The proposed topology's switching waveforms for time interval between t0 and t1

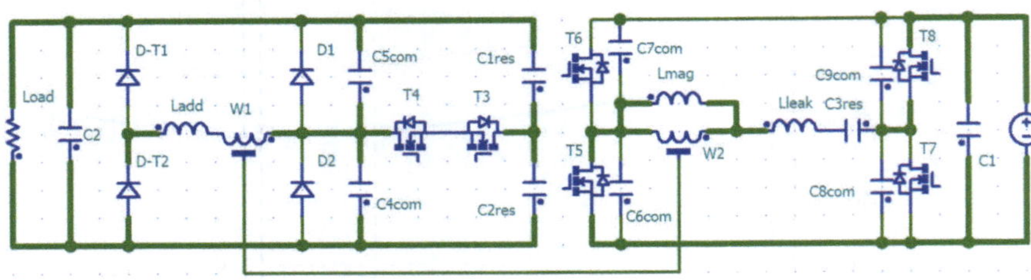

(**b**) The proposed topology's switching waveforms for time interval between t1 and t2

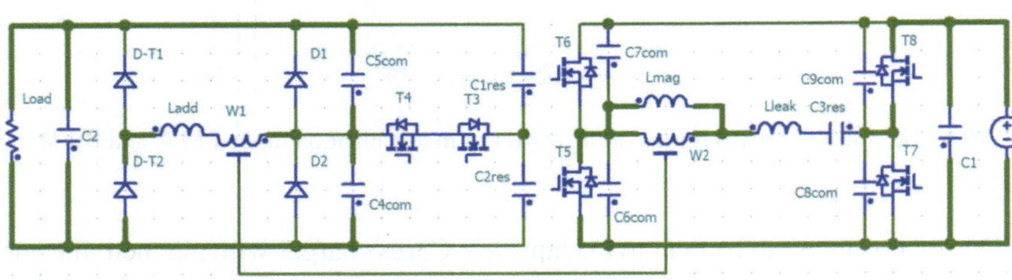

(**c**) The proposed topology's switching waveforms for time interval between t2 and t3

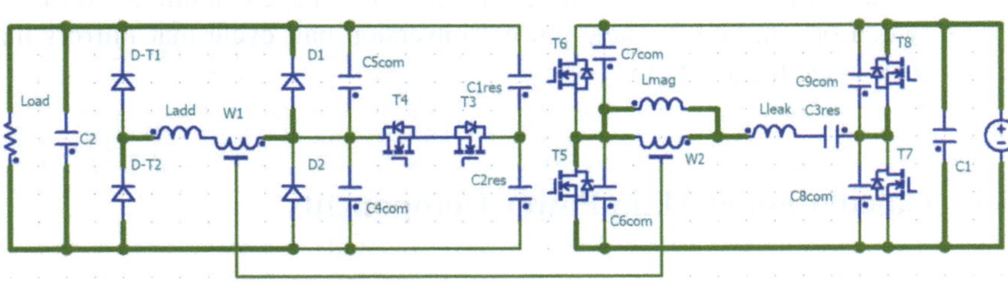

(**d**) The proposed topology's switching waveforms for time interval between t3 and t4

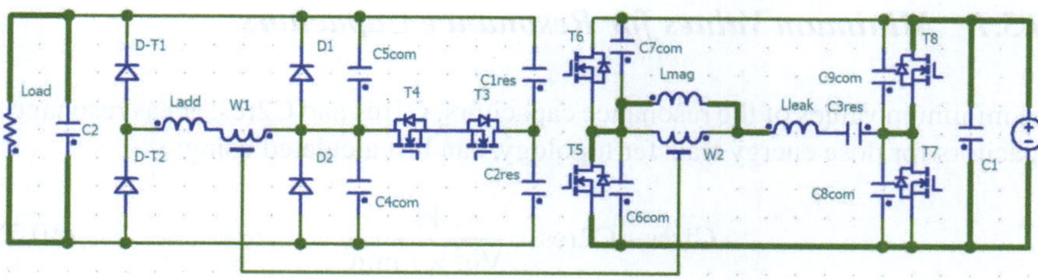

(e) The proposed topology's switching waveforms for time interval between t4 and t5

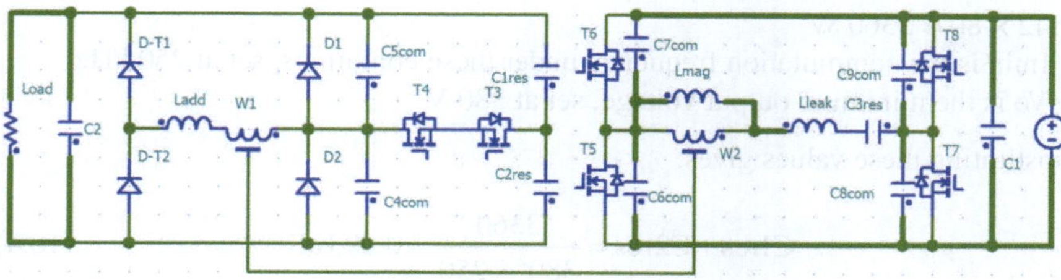

(f) Switching waveforms for the time interval between t5 and t6

The transformer ratio (*N*) is given by:

$$N = Vh / 2Vl = 400 / 2 \times 50 = 4 \tag{10.1}$$

where

- V_h is the middle voltage on the high-voltage side
- V_l is the middle voltage on the low-voltage side

To verify the turn ratio of the transformer using another method:

$$N = \text{Vin} / 2\text{Vout} \times \text{Mmin} = 400 / 2 \times 50 \times = 3.81 \tag{10.2}$$

where the gain Mmin $= \sqrt{\dfrac{m}{m-1}} = \sqrt{\dfrac{11}{11-1}} = 1.05$

Finally, the selected turn ratio of the transformer is 4, as a turn ratio of 3.81 would be challenging to implement. During simulation calculations, we must also pay attention to the step-down mode, where we need to obtain a 60 V output.

The preliminary leakage inductance of the transformer is 52 nH. This value is a summation of the leakage inductance of the power transformer itself (23 nH) plus the inductance from the PCB and components on the low-voltage side (27 nH).

10.5.1 Minimum Values for Resonance Capacitors

The minimum values of the resonance capacitors, C1res and C2res, act as resonance capacitors for dose energy transfer topology, can be calculated using:

$$\mathrm{C1res} + \mathrm{C2res} = \frac{\mathrm{Pl}}{\mathrm{Vo}^2 \times f\,\mathrm{min}} \tag{10.3}$$

where

- Pl is the maximum output power, which at minimum input voltage is $42 \times 80 = 3360$ W
- fmin is the commutation frequency under these conditions, set at 250 kHz
- Vo is the minimum output voltage, set at 380 V

Substituting these values gives:

$$\mathrm{C1res} + \mathrm{C2res} = \frac{3360}{380^2 \times 250} = 0.093\,\mathrm{uF} \tag{10.4}$$

Next, we calculate the maximum current in the worst-case scenario through these capacitors. Since C1res is equal to C2res, the current through these capacitors is half of the current flowing through the inductor Ladd:

$$Ia = \pi \times \mathrm{Io} \frac{\mathrm{fres}}{\mathrm{fmin}} \sqrt{Ds/2} = 23.67\,\mathrm{A} \tag{10.5}$$

where

$$\mathrm{Io} = 3360/380 = 8.84\,\mathrm{A}$$

fres is the resonance frequency, calculated as 364 kHz

- Ds is the duty cycle of the pulse current when the commutation frequency is at its minimum (250 kHz), approximately 0.685.

We select 10p of SMA capacitors 10 nF in package 1812 with NPO material. The dissipation factor of NPO material in the worst-case conditions is 0.1%, leading to an equivalent series resistance (ESR) of 0.064 ohms. The maximum power dissipation for the 1812 package is 0.4 W. Considering derating guidelines for components, the current through one capacitor is approximately 2.4 A, resulting in a dissipation power of: Pdiss = 0.368 W. This is too high, so we select 12p capacitors, which yield a dissipation power of 0.256 W, making it more acceptable.

The final values of the capacitors, C1res and C2res, will be 60 nF each. The important note. It is important to follow a rule that all parts must have a minimum of two sources. When selecting components, it's essential to follow the rule that all parts must have at least two sources. This ensures that if one part becomes unavailable, an alternative can be used, preventing production delays. In practice, reviewing data from different companies indicates that the worst-case ESR is around 0.048 ohms. In this case, power dissipation would be: Pdiss = 0.192 W.

Capacitors with a reasonable cost typically have an ESR of 0.013 ohms, leading to power dissipation of: Pdiss = 0.052 W.

The capacitor C3res is calculated using the following equation:

$$\text{C3res} = {}^{1}\!\big/_{2\pi \times Q \times \text{fres} \times \text{Rac}} = 1.96\,\text{uF} \tag{10.6}$$

where

$$\text{Rac} = {}^{\text{Vout}}\!\big/_{2 \times 16 \times 24} = 0.495\,\Omega$$

- $Q = 0.45$ (chosen based on normalized graphics with $m \approx 10$–11).

The selected capacitor is 2 μF. The current through this capacitor is 24 A × 4 = 96 A. We choose 20 capacitors, each rated at 100 nF and 200 V in package 1812 with NPO material and an ESR of 0.0032 ohms. The current through each capacitor will be: 96/20 = 4.8 A.

For calculations, we will use 5 A. The dissipation power for each capacitor is: Pdiss = 0.08 W.

This value is acceptable.

10.5.2 Calculation of Additional Inductor

The leakage inductance of transformer is 52 nH. We cannot increase this inductance, as it would significantly increase power losses and costs. The maximum commutation frequency is 340 kHz. In this case, the resonance frequency is 364 kHz.

The total resonance capacitor is: 2/16 × 0.12/(2/16 + 0.12) = 0.0612 Uf.

The resonance inductor is calculated as: Lres = 1/39.4 × 0.0612 × 364^2 = 3.125 uH.

The inductance of transformer reflected on high voltage side is 0.053 × 16 = 0.832 uH.

And the additional inductor Ladd is: 3.125–0.832 = 2.29 uH.

We selected power transistors and performed simulations/calculations of the converter in both step-up and step-down modes under various input voltage conditions. Figure 10.3 shows the schematic of the converter, where transistors T1–T2 are UJ4SC075011B7S, T3–T4 are UJ4SC075009B7S, and T5–T8 are 3p FDBL86063-F085 connected in parallel. Capacitor C1 is 60 μF and C2 is 2 μF. The commutation capacitors C6–C9 are each 20 nF. The conductive losses of the PCB connections on the low-voltage side were calculated to be around 1.1 mΩ. Other losses from the PCB and filter are not included in this calculation. Tables 10.1 and 10.2 show the results of simulations/calculations in step-up mode with limited input current of around 80 A.

Table 10.3 shows the results in step-up mode with different low loads. Tables 10.4 and 10.5 display the results of simulations/calculations in step-down mode with a maximum output current of around 40 A. In this mode, it is important to monitor the output voltage to ensure the battery is fully charged without exceeding the charge limit.

Overcharging is completely unacceptable for lithium-ion batteries; therefore, we performed simulations to reduce the duration of the pulses for T1 and T2. Table 10.6

Table 10.1 Conductive losses of main components and total losses when step-up to Vout = 380-400 V

Vin	Pout	f com	T1–T2	T3–T4	T5–T8	C3res	Tr	Total	Eff	P-C
40 V	3198 W	277 kHz	19 W	11 W	20 W	10 W	26 W	86 W	0.974	72%
46 V	4106 W	340 kHz	23 W	14 W	23 W	12 W	30 W	103 W	0.976	91%
58 V	4589 W	340 kHz	20 W	8 W	18 W	9.1 W	27 W	82.1 W	0.982	80%

Table 10.2 Switching stress and conductive stress main power components

Vin	Ioff-T5-T8	I-C2 RMS	I-C2res	I-C4com	Tr-VA	Pout/Tr VA	I-LV Tr
40 V	67 A	6.1 A	10.3 A	1.3 A	4246	0.75	98.3 A
46 V	51 A	6.2 A	11.6 A	0.5 A	5136	0.8	107 A
58 V	58 A	7.7 A	10 A	1.7 A	5575	0.82	95.3 A

Table 10.3 Different power levels with different input voltages

	Around 50% load		Around 10% load	
Vin	f com	Eff.	f com	Eff.
40 V	277 kHz	97.8%	277 kHz	92.7%
46 V	340 kHz	97.7%	340 kHz	95.0%
58 V	340 kHz	97.8%	340 kHz	90.0%

Table 10.4 Conductive losses of main components and total losses step-down Vin = 410 V

Vout	Pout	f com	T1–T2	T3–T4	T5–T8	C3res	Tr	Total	Eff	P-C
40 V	1600 W	340 kHz	2.5 W	5 W	6 W	3 W	18 W	34.5 W	97.9%	68%
49 V	2113 W	340 kHz	2 W	7 W	6 W	3 W	17 W	35 W	98.4%	82%
59.5 V	2355 W	250 kHz	3 W	9 W	6 W	3 W	20 W	41 W	98.3%	60%

Table 10.5 Switching stress and conductive stress main power components

Vin	I-C1 RMS	I-C2res	I-C4com	Tr-VA	Pout/Tr VA	I-LV Tr
40 V	31 A	7 A	1.9 A	1988	0.8	52.5 A
46 V	30.5 A	6.2 A	0.5 A	2654	0.8	53 A
58 V	38.6 A	7.5 A	0.8 A	3391	0.69	55.5 A

Table 10.6 Step-down mode characteristics with a reduced duration of the pulses for T1 and T2 for light load operation

Vout	f com	Pulse T1–T2	Pout	Eff
60 V	250 kHz	1.2 uS	245 W	78.6%

Table 10.7 Characteristics of step-up mode when the output voltage is 395 V

Vin	41.2 V	43 V	51.4 V	58 V
Pout	3195 W	3583 W	3949 W	4054 W
Comm. f	250 kHz	270 kHz	340 kHz	340 kHz
Efficiency	0.955	0.956	0.97	0.96

Table 10.8 Characteristics of step down mode when the input voltage is 405 V

Pout	1790 W	2190 W	1700 W	1543 W
Vout	42.3 V	46.8 V	55 V	58 V
Comm. f	334 kHz	333 kHz	238 kHz	238 kHz
Efficiency	96.0%	97.5%	96.4%	95.7%

shows results in step-down mode with a reduced duration of the pulses for T1 and T2 at very low load.

Upon analyzing all results, we can see that a turn ratio of 4 for the power transformer is not optimal but is acceptable. A turn ratio of 3.81 would be difficult and expensive to implement. Overall, all results are acceptable, leading us to the decision to build a prototype.

It is important to note a very practical engineering point: if you have acceptable results, move forward. Seeking the best results can consume more time without guaranteeing that the outcome will improve. In other words, engineering is an art of compromise.

10.6 Experimental Results

Table 10.7 presents the results in step-up mode, while Table 10.8 shows the results in step-down mode. The converter's losses include conductive losses on both the low-voltage and high-voltage sides, as well as the EMI filters. Figures 10.6, 10.7, and 10.8 display oscillograms under various conditions in step-up mode. On the high-voltage side in step-up mode, synchronous rectification is used for transistors T1 and T2. This approach helps reduce losses by up to 15 W. Figures 10.9, 10.10, and 10.11 provide oscillograms under various conditions in step-down mode.

Analyzing the data from the test prototype, we can conclude that the results obtained during testing are acceptable. The prediction regarding the issue with the power transformer's turn ratio in step-up mode was validated. Achieving a 40 A

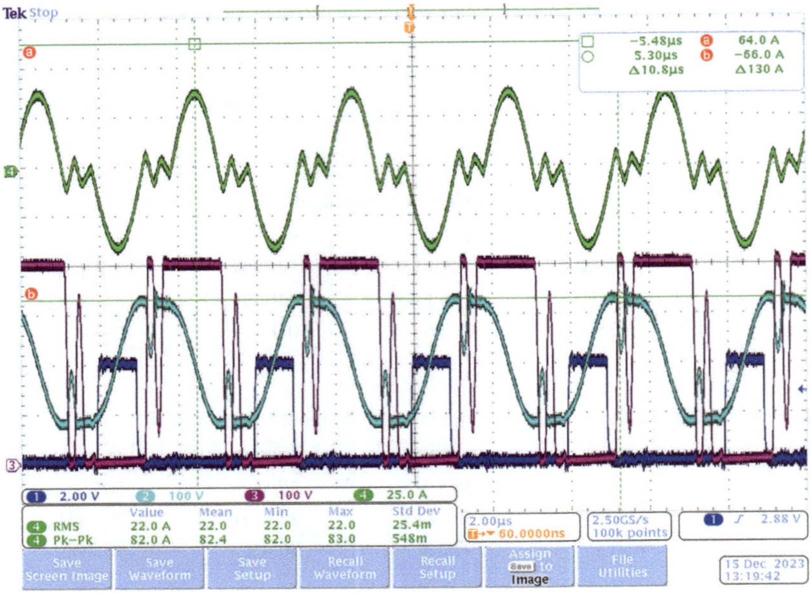

(a) High Voltage Side

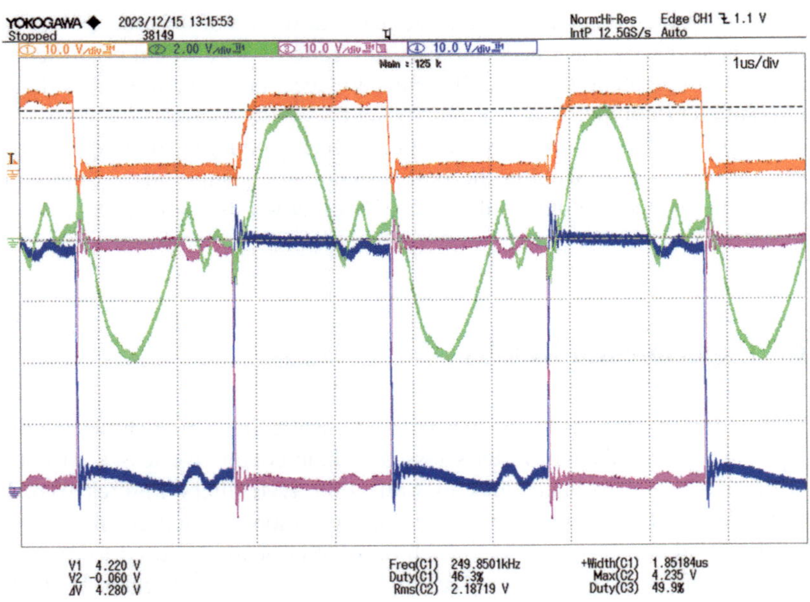

(b) Low Voltage Side

Fig. 10.6 Step-up mode: V_{IN} = 41 V. High voltage side: Ch.1-gate T1, Ch.2-C4com, Ch.3-T1, Ch.4-Current Ladd. Low voltage side: Ch.1-gate T5, Ch.2-CurrentC3res, Ch.3-T5, Ch.4-T7. (**a**) High voltage side. (**b**) Low voltage side

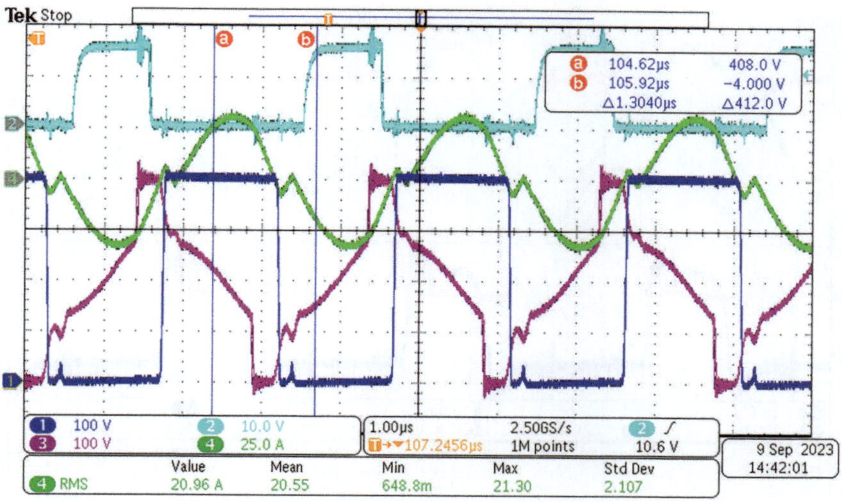

(a) High Voltage Side

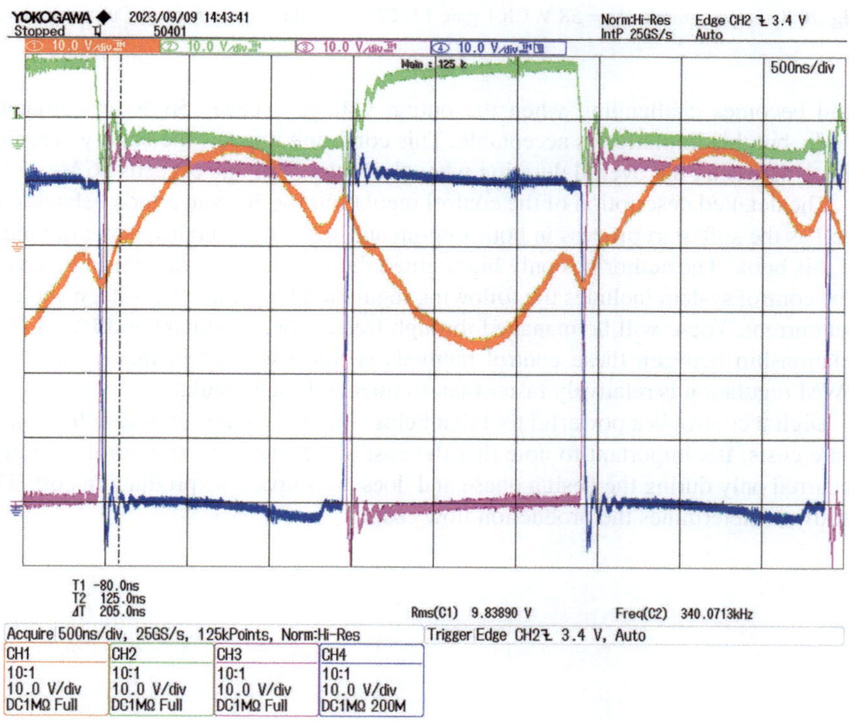

(b) Low Voltage Side

Fig. 10.7 Step-up mode V_{IN} = 52 V. High voltage side: Ch.1-T1, Ch.2-gate T1, Ch.3-C4com, Ch.4-Current Ladd. Low voltage side: Ch.1-Current C3res, Ch.2-gate T5, Ch.3-T5, Ch.4-T7. (**a**) High voltage side. (**b**) Low voltage side

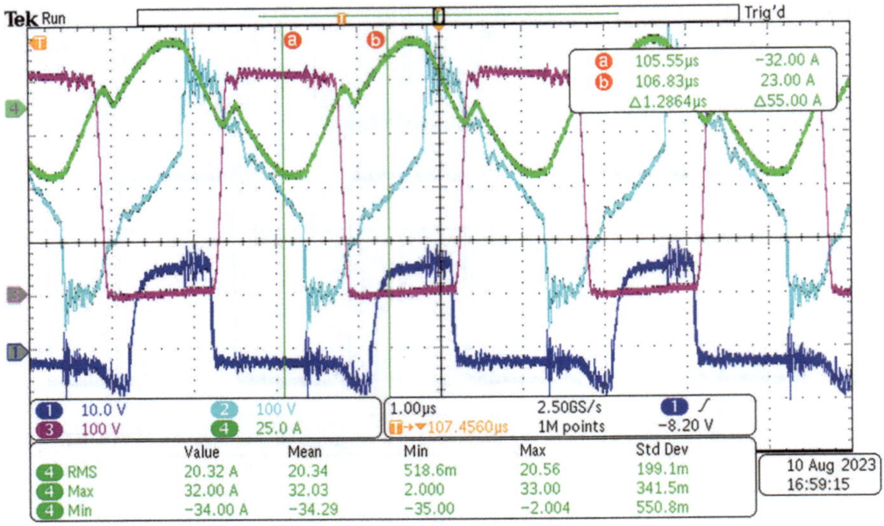

Fig. 10.8 Step-up mode $V_{IN} = 58$ V. Ch.1-gate T1, Ch.2-C4com, Ch.3-T1, Ch.4-Current Ladd

load becomes challenging when the output voltage exceeds 56 V in step-down mode, but this limitation is acceptable. This condition impacts the battery charging time, increasing the overall duration when the battery voltage exceeds 56 V.

The detailed description of the control regulation for the converter's behavior, as well as the soft-start process in both step-up and step-down modes, is not the focus of this book. The author has only highlighted the main aspects of control behavior. The control system includes the following regulation loops: output voltage and output current. These will be managed through frequency adjustment and PWM. The relationship between these control methods is not discussed in this book. Note: PWM regulation is relatively faster than frequency-based regulation.

Digital control is a powerful tool that helps simplify the design and reduce hardware costs. It's important to note that the cost of developing the control system is incurred only during the design phase and does not impact the production cost. The hardware determines the production flow costs.

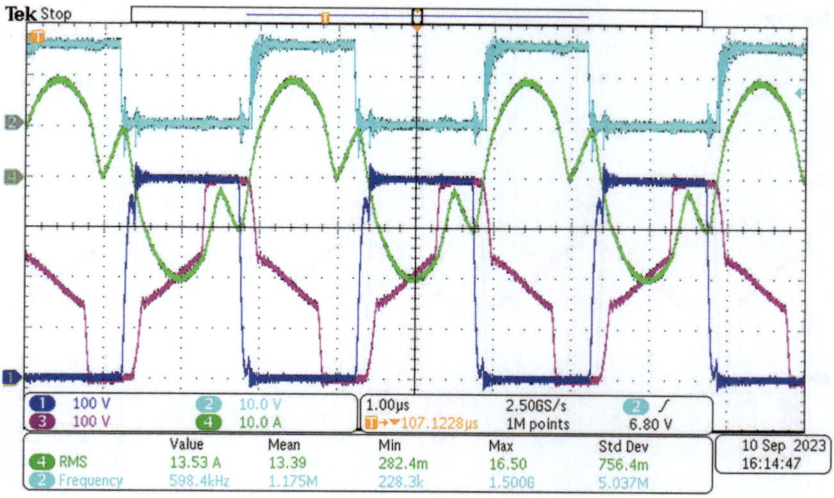

(a) High Voltage Side

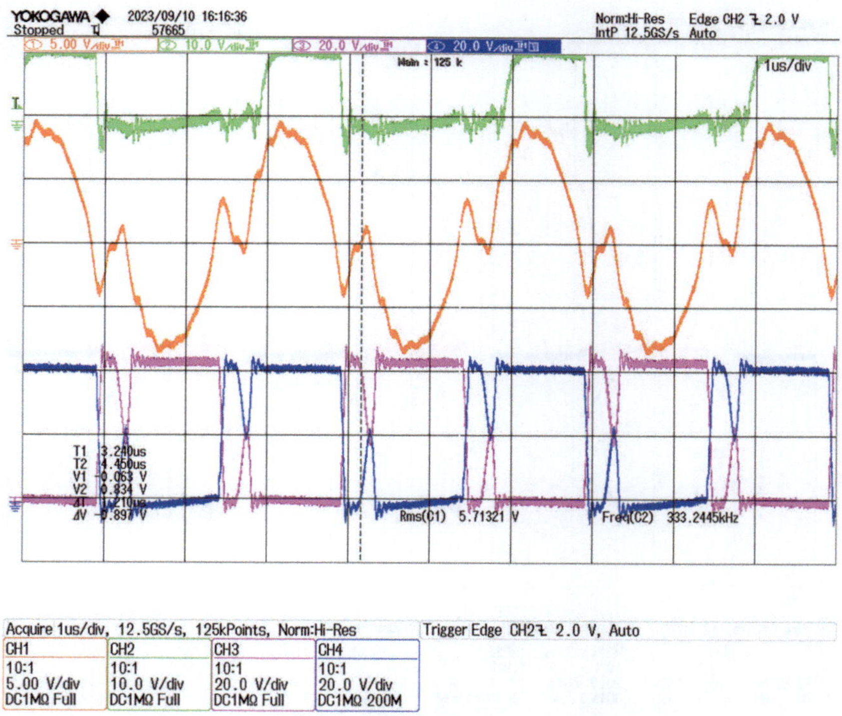

(b) Low Voltage Side

Fig. 10.9 Step-down mode $V_{OUT} = 42$ V. High voltage side: Ch.1-T1, Ch.2-gate T1, Ch.3-C4com, Ch.4-Current Ladd. Low voltage side: Ch.1-Current C3res, Ch.2-gate T5, Ch.3-T5, Ch.4-T7. (**a**) High voltage side. (**b**) Low voltage side

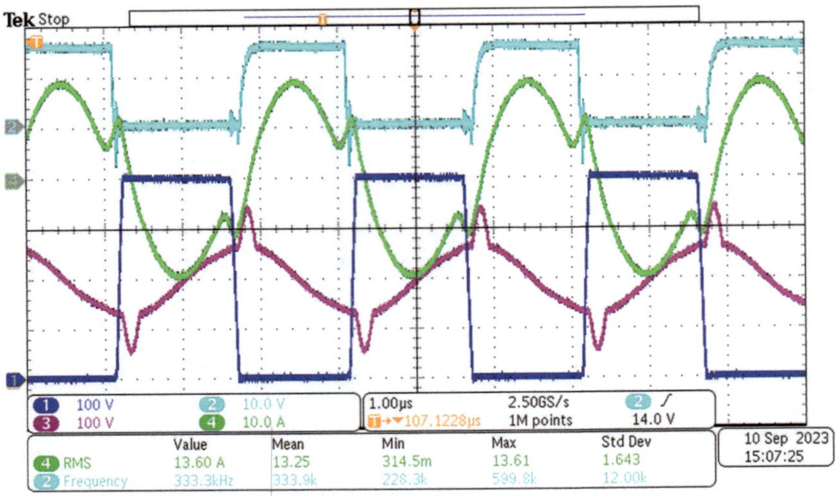

(a) High Voltage Side

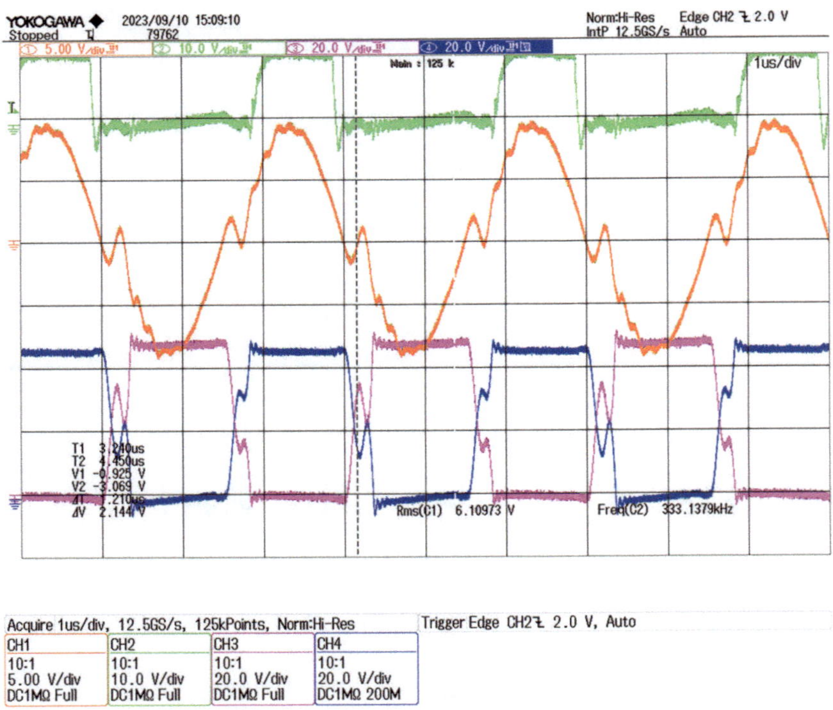

(b) Low Voltage Side

Fig. 10.10 Step-down mode $V_{OUT} = 47$ V. High voltage side: Ch.1-T1, Ch.2-gate T1, Ch.3-C4com, Ch.4-Current Ladd. Low voltage side: Ch.1-Current C3res, Ch.2-gate T5, Ch.3-T5, Ch.4-T7. (**a**) High voltage side. (**b**) Low voltage side

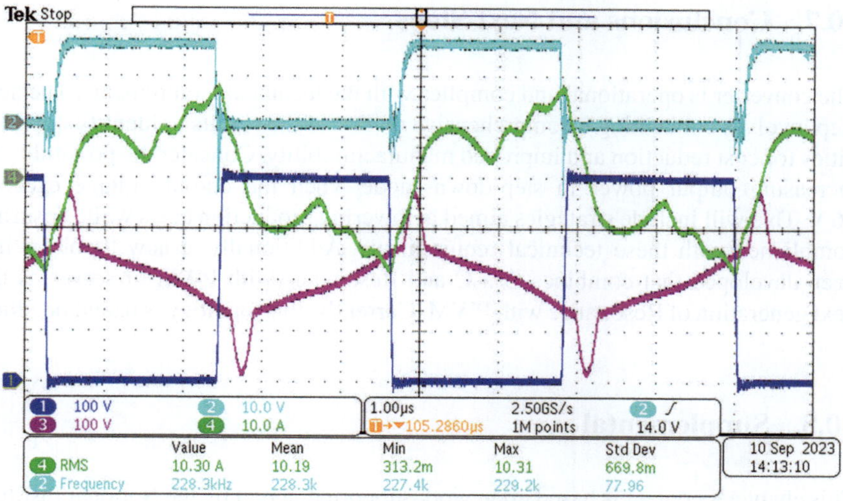

(a) High Voltage Side

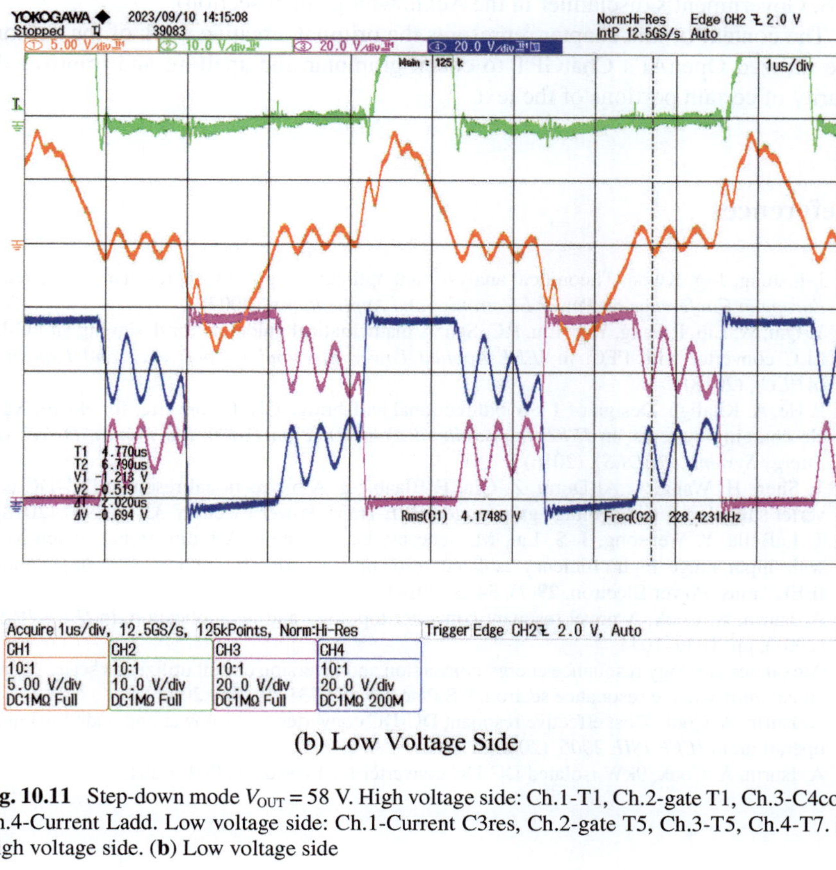

(b) Low Voltage Side

Fig. 10.11 Step-down mode $V_{OUT} = 58$ V. High voltage side: Ch.1-T1, Ch.2-gate T1, Ch.3-C4com, Ch.4-Current Ladd. Low voltage side: Ch.1-Current C3res, Ch.2-gate T5, Ch.3-T5, Ch.4-T7. (**a**) High voltage side. (**b**) Low voltage side

10.7 Conclusions and Next Steps

The converter is operational and complies with the technical requirements. The next step involves conducting a comprehensive review of the results to identify opportunities for cost reduction and improved manufacturability. Consider the possibility of increasing output power in step-down mode when the output voltage exceeds 56 V. This will include strategies aimed at lowering production costs while ensuring compliance with these technical requirements. Additionally, a new topology has been developed that combines CLLC and Resonance with PWM or serves as the next generation of Resonance with PWM. Currently, this topology is patent pending.

10.8 Supplemental

This chapter's content is based upon work supported in part by the National Institute for Standards and Technology under Award Number 70NANB22H007 (see U.S. Government's disclaimer in the Acknowledgement section).

The content of this chapter represents the original, creative work of the authors. We utilized OpenAI's ChatGPT to check grammar and spelling and improve the clarity of certain portions of the text.

References

1. J.-h. Jung, J.-g. Kwon, Theoretical analysis and optimal design of LLC resonant converter, in *European Conference on Power Electronics and Applications*, (2007)
2. Y. Qiu, W. Liu, P. Fang, Y.-F. Liu, P.C. Sen, A mathematical guideline for designing an AC-DC LLC converter with PFC, in *IEEE Applied Power Electronics Conference and Exposition (APEC)*, (2018)
3. P. He, A. Khaligh, Design of 1 kW bidirectional half-bridge CLLC converter for electric vehicle charging systems, in *IEEE International Conference on Power Electronics, Drives and Energy Systems (PEDES)*, (2016)
4. Y. Shen, H. Wang, A. Al-Durra, Z. Qin, F. Blaabjerg, A bidirectional resonant DC–DC converter suitable for wide voltage gain range. IEEE Trans. Power Electron. **33**(4), 2957 (2018)
5. T. LaBella, Y. Wensong, J.-S. Lai, M. Senesky, D. Anderson, A bidirectional-switch-based wide-input range high-efficiency isolated resonant converter for photovoltaic applications. IEEE Trans. Power Electron. **29**(7), 3473 (2014)
6. A. Isurin, A. Cook, A novel resonant converter topology and its application, in *IEEE PESC*, (2001), pp. 1039–1044
7. Alexander topology resonance energy conversion and inversion circuit utilizing a series capacitance multi-voltage resonance section, US Patent 6,483,731, 19 Nov 2002
8. A. Isurin, A. Cook, Cost effective resonant DC-DC converter for hi-power and wide load range operation, in *IEEE ISIE 2006*, (2006)
9. A. Isurin, A. Cook, 9kW isolated DC-DC converter for hybrid bus PCIM 2012

Chapter 11
Conclusions

11.1 Advancements in High-Power DC–DC Converters: Embracing Wide Bandgap Semiconductors and Soft-Switching Technology

In previous chapters, we explored various authors' perspectives on implementing high-power, high-current isolation DC–DC converters, particularly those exceeding 200 A. The authors hope this book will assist design engineers in creating such converters. While it is impossible to write a book that provides direct instructions for every scenario—since real-world conditions constantly bring new challenges and requirements—the goal is to anticipate and address as many potential issues as possible.

Life is constantly evolving, and so is the field of power electronics and its related technologies. We observe many interesting trends, particularly in transportation, renewable energy, industry, and other areas. These trends are often supported by new technologies, which either help create new trends or enable the realization of previously unattainable innovations.

The development of power electronics is well illustrated by the impact of new technologies. When design engineers gained access to MOSFETs and new types of passive components, it sparked a revolution in the field. Additionally, advancements in microelectronics played a crucial role. These technologies are deeply interconnected, and their evolution has raised many collateral questions and challenges, which have been successfully addressed over time.

Currently, we are witnessing the next step in the development of power electronics: the introduction of wide bandgap (WBG) transistors. These transistors offer excellent performance characteristics and are becoming more cost-effective. WBG semiconductors are highly promising, allowing design engineers to create converters with significantly improved performance, both in terms of cost and power density.

A. Isurin, M. J. Scott, *Practical Design Considerations for Isolated DC-DC Converters*, https://doi.org/10.1007/978-3-031-94427-7_11

The authors had the opportunity to validate this statement. In the DC–DC converter described in the article "9 kW Isolated DC–DC converter for Hybrid Bus," silicon (Si) transistors were replaced with silicon carbide (SiC) transistors. This substitution improved efficiency by 3.4% at 9 kW, increased the power capacity to 13.2 kW, and enhanced efficiency by an additional 2.4%. Additionally, the power stage cost was reduced by 10%, and the switching frequency was increased. It's important to note that these improvements were possible because the original converter utilized a pure soft-switching technique.

The authors hold a strong opinion that wide bandgap (WBG) transistors should be used in soft-switching operation mode. Only under this condition can design engineers fully realize the advantages of WBG technology, ensuring good reliability and cost-effectiveness. Therefore, employing soft-switching is essential for maximizing the benefits of WBG transistors.

Over the last 20 years, there has been a growing demand for improved electromagnetic interference (EMI) performance, longer lifetimes, and minimal maintenance for all types of power converters, and this trend is continuing. The combination of WBG semiconductors and soft-switching technology helps achieve these goals, leading to significant advancements in performance and reliability.

11.2 Definition of an Ideal Isolated DC–DC Converter

Cost-effectiveness and superior performance are key targets for design engineers and represent the final goal. This book focuses on isolated DC–DC converters, and now we aim to finalize the definitions, tools, recommendations, and targets for designing these converters. In other words, we will attempt to define the characteristics of the ideal isolated DC–DC converter, specifically in the context of high-power and high-current applications.

The application area extends to the continuation of the operation area, so we will begin by examining the operational characteristics. The main features required for an ideal power converter include:

- **Soft-switching** across the entire power and input/output voltage range
- **Controlled energy transfer** using dosed energy techniques
- **High duty cycle**, with power conversion efficiency maintained at least 90% across the full power and voltage range
- **Regulation** achieved through pulse-width modulation (PWM) under full power conditions
- **Zero voltage,** a power converter able to provide an output voltage from zero volts to the nominal output voltage when the converter is running
- **Split current paths** to prevent interrupted high-level current on the low-voltage and high-current side
- **Low interwinding capacitance** in the power transformer
- **Simplified magnetic components** for easier design and maintenance

- **Minimal use of active switches** to enhance reliability and efficiency
- **Avoidance of auxiliary circuits** wherever possible to streamline the design

These features define the basic requirements of an ideal converter. Let's delve into each aspect in detail.

1. **Soft switching** provides lower switching losses, resulting in higher efficiency, lower heat dissipation, smaller cooling systems, and reduced electromagnetic interference (EMI). It also produces lower stresses on converter components, leading to longer lifetimes and higher reliability. With soft switching, a higher commutation frequency can be achieved, resulting in smaller passive components, lower output ripple, and better dynamic response.

To implement soft switching effectively, follow these guidelines:

   - The switch that operates with the full bus voltage should interrupt a current level of approximately 10% of the peak current. This current level should remain relatively consistent across the full voltage and power range. With this current, achieving Zero Voltage Switching (ZVS) turn-off is straightforward. The turn-on for this switch should be under both Zero Current Switching (ZCS) and ZVS conditions. The Resonant with PWM topology illustrates this rule.

     The switch that interrupts relatively high current needs to turn off under pure ZVS conditions. The commutation voltage level should follow the rule: higher current corresponds to lower voltage. In other words, this switch should never commutate the full bus voltage. This is illustrated by the horizontal switch in the Resonant with PWM topology and the main switches in the dual current-fed topology with switched capacitors.

2. **The** primary advantage of power conversion topologies that utilize dosed energy transfer is their ability to self-adjust under overload conditions, transitioning from a voltage source mode to a power source mode without external intervention. This self-regulation prevents overloading of the power source, enhancing the reliability of the converter compared to systems that require external control for overload protection. These converters are particularly well-suited for applications where multiple power converters are connected in parallel and where the load demands consistent power sourcing. Examples of such topologies include the resonant topology with PWM and the dual current-fed topology with switched capacitors.
3. The **coefficient of power conversion** indicates how efficiently the semiconductors and power transformer are utilized. Optimal power conversion occurs when the current flow is interrupted during the dead time of active power semiconductors. This practice reduces the current stress on the DC-link capacitors, allowing for smaller capacitors to be used, which in turn reduces the overall size and cost of the converter. Additionally, a higher coefficient of power conversion results in reduced electromagnetic interference (EMI).
4. **Regulating the output** voltage or current by changing the commutation frequency conflicts with achieving a high coefficient of power conversion.

Therefore, **regulation** should be accomplished by **PWM** at the maximum commutation frequency employed by the power converter.

5. The ability of a power converter to **output zero volts** while running and under control. The main **reasons** are that it enables the converter to easily soft-start and allows it to protect itself under an overload condition.
6. **Split current paths**. For the low-voltage, high-current side, using a dual current-fed **configuration** is advantageous. In this setup, the active switches on the low-voltage side should turn on and off under zero-current switching (ZCS) conditions, ensuring that high currents are never interrupted. This eliminates the need for snubber circuits and large DC-link capacitors. If ZCS cannot be implemented, a simpler approach with larger block capacitors can be used.
7. The **power** transformer must have **interwinding capacitance** as small as possible, ideally in the range of several picofarads (pF). Excessive interwinding capacitance can cause electromagnetic interference (EMI) and affect loop stability. Due to its significant impact, this factor is highlighted separately.
8. **The size of magnetic components** should be minimized, and the commutation frequency should be as high as possible. These measures help reduce the size, cost, and response time of the control loop.
9. **Active switches** significantly impact the cost, control complexity, and reliability of the **converter**. They are also a source of EMI. For high-power applications above 2 kW, a minimum of four active switches (excluding synchronous rectification) is typically required, and up to eight for bi-directional isolation DC–DC converters. For converters up to 1 kW, two active switches for unidirectional and four for bi-directional configurations are generally the maximum. The decision between one and two switches is often debated in terms of efficiency and practicality.
10. **Auxiliary circuits**, such as capacitors connected in parallel to active switches, are sometimes necessary but can be controversial due to their impact on overall design complexity and **performance**. Passive energy recovery snubbers can be effective for handling commutation currents up to 100 A, but where possible, it is preferable to minimize the use of auxiliary circuits to simplify the system.

The application area for an ideal converter requires it to be bi-directional, with an input voltage range that spans twice the nominal voltage and an output voltage range that varies by two to three times. The converter should maintain consistent performance across all combinations of input and output voltages, even as the load varies from 10 to 100%—ideally, from 0 to 100%. Currently, topologies such as resonant converters with PWM and dual current-fed converters with switched capacitors closely approximate the characteristics of an ideal DC–DC converter.

11.3 Recommendation and Target to Strive When Designing Power Converter

It is evident that the views expressed here represent the opinions of the authors. However, we strongly believe that design engineers need to clearly understand the targets they are striving for in order to achieve truly excellent products. We invite everyone to join in this ongoing discussion to refine and clarify the definition of the ideal DC–DC converter.

It is worth noting that the nine points outlined for the operation area of a DC–DC converter are broadly applicable to any type of power converter.

Appendix

Measuring Voltage and Current in a System

To ensure the proper functionality of a power converter, it is essential to control and measure voltage and current at both the input and output, along with various other parameters. The primary parameters are voltage and current. Let's start with voltage.

While every electrical engineer knows how to measure voltage, it is important to highlight several practical considerations specific to DC–DC converters. The measurement circuits used for the control loop must be connected directly to the bulk capacitor, positioned before the filter, and never after. The filter can interfere with the phase and frequency characteristics of the control loop, potentially causing stability issues. The measuring circuit must have as low an impedance as possible. This allows for a relatively easy provision of a higher corner frequency in the measuring circuit. The precision of the measuring circuit is a critical aspect of the technical specifications of the converter. However, this is particularly crucial when the DC–DC converter is connected to a battery. In such cases, the battery voltage must be measured with a maximum tolerance of 0.1%, and the ripple voltage, peak-to-peak, provided by the converter to the battery should be around 0.5%. The ripple refers to low-frequency fluctuations, typically several hertz (1–10 Hz), which occur when using digital control.

The next point is the practical implementation of voltage measuring circuits suitable for DC–DC converters, which we consider in Chap. 6. Figure A.1 shows the basic schematic of measuring voltage on the low-voltage side of a DC–DC converter, both without offset and with offset. The operational amplifier (Op Amp) is powered by a 5 V supply. The output signal from the Op Amp can be connected directly to the control board. A 4 V reference is used as a precise voltage source with a tolerance of +/−0.025%.

The difference between these two circuits is that the output voltage from the Op Amp with offset starts to rise from zero when the input voltage reaches around

A. Isurin, M. J. Scott, *Practical Design Considerations for Isolated DC-DC Converters*, https://doi.org/10.1007/978-3-031-94427-7

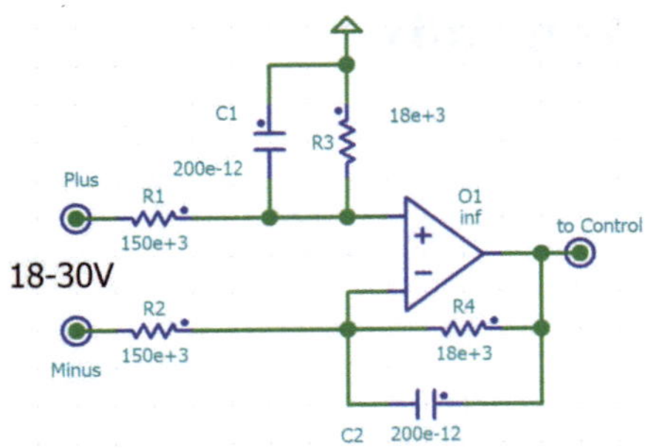

Fig. A.1 The basic schematic of measuring voltage on the low-voltage side of a DC–DC converter (a) without offset

(b) with offset

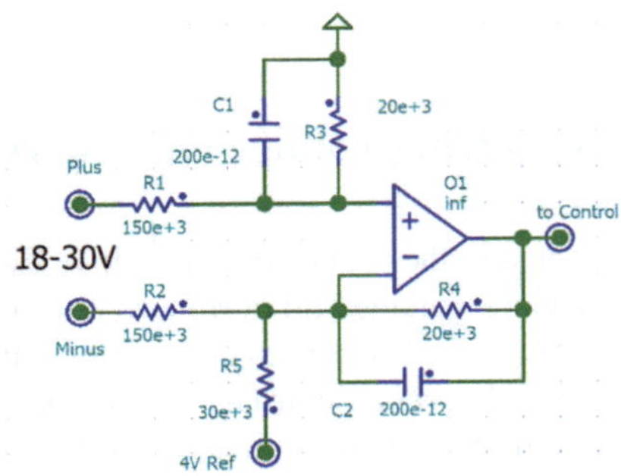

13 V. In other words, the Op Amp with offset effectively measures input voltages from 13 to 30 V, whereas the Op Amp without offset measures from 0 to 30 V. This means that the resolution of the measuring circuit with offset is twice as high as that of the circuit without offset. This feature significantly helps achieve stable control. Note that it is generally unnecessary to measure voltages below 50% of the nominal voltage.

Figure A.2 shows the basic schematic of measuring voltage on the high-voltage side of a DC–DC converter, both without offset and with offset. The Op Amp and voltage reference are the same as in Fig. A.1. The Op Amp is connected to an isolated amplifier, which is then connected to the control board. In this case, the behavior is the same as on the low-voltage side, and the final result is that the resolution of the measuring circuit with offset is twice as high as that of the circuit without offset.

Any power converter requires the ability to measure and control the current flowing through the power stage. The measurement data is used to control the converter and provide protection from overcurrent, preventing fatal damage. Isolation DC–DC converters, in particular, have specific requirements for these circuits. One key requirement is that the current sensor must be placed on the primary side of the transformer (before the power transformer). This placement ensures that if something happens to the power transformer or the secondary side (such as an overload or short circuit in the output circuit), the current sensor will detect it and control the system to prevent fatal damage. In other words, the current sensor is located on the

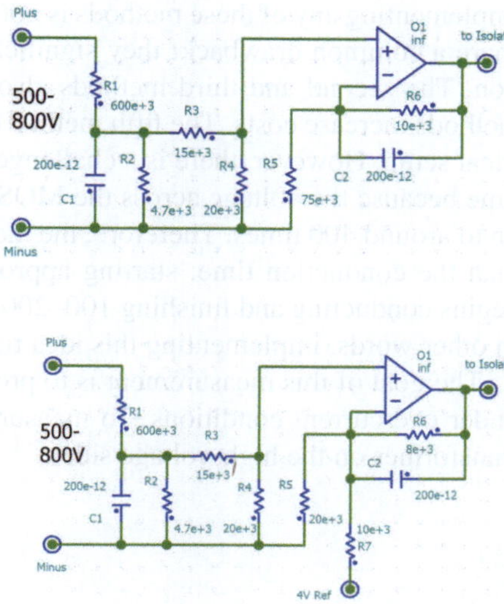

Fig. A.2 the basic schematic of measuring voltage on the high-voltage side of a DC–DC converter (**a**) without offset

(**b**) with offset

low-voltage side in a step-up converter and on the high-voltage side in a step-down converter. In a bi-directional converter, a current sensor is needed on both sides of the power transformer.

The current sensor circuit must have a fast response time. The total time from detecting an anomaly to shutting down the converter or starting current limiting cannot exceed 0.5 μs.

For the high-voltage side, a current transformer can be used for protection and for measuring current. This transformer can handle currents up to 200 A and is suitable for both input and output current measurements in step-up and step-down modes.

On the low-voltage side, the situation is quite different. Measuring the current through the low-voltage winding of a power transformer, especially when dealing with high currents up to 1000 A, presents significant challenges. There are several potential solutions to this problem:

1. **Special current transformer**: Using a current transformer specifically designed to handle high currents
2. **Split low-voltage winding**: Dividing the low-voltage winding so that the main current (90–95%) flows directly through the winding, while a smaller portion (5–10%) passes through a standard current transformer
3. **Hall sensor with closed loop and split current**: Implementing a Hall sensor in a closed-loop configuration, with split current, as Hall sensors for 1000 A are typically too large
4. **Rogowski coil**: Utilizing a Rogowski coil to measure the current
5. **Measuring the voltage across the MOSFET**: Measuring the voltage drop across the MOSFET when it is conducting to determine the on-state resistance (Rds(on)) of the MOSFETs

Implementing any of these methods is not straightforward. The first four approaches share a common drawback: they significantly impact the mechanical implementation. The second and third methods also affect electrical conditions, and all four methods increase costs. The fifth method does not influence the mechanical or electrical setup. However, there is a challenge in measuring only during the conduction time because the voltage across the MOSFET varies significantly, with differences up to around 400 times. Therefore, the measurement duration must be much shorter than the conduction time: starting approximately 100–200 ns after the MOSFET begins conducting and finishing 100–200 ns before the MOSFET stops conducting. In other words, implementing this idea requires a specialized schematic design.

The goal of this measurement is to provide protection by shutting down the unit under overcurrent conditions. To measure current, it is necessary to use a current transformer on the high-voltage side.

Index

A. Isurin, M. J. Scott, *Practical Design Considerations for Isolated DC-DC Converters*, https://doi.org/10.1007/978-3-031-94427-7

MIX
Papier aus verantwortungsvollen Quellen
Paper from responsible sources
FSC® C105338

If you have any concerns about our products, you can contact us on
ProductSafety@springernature.com

In case Publisher is established outside the EU, the EU authorized representative is:
Springer Nature Customer Service Center GmbH
Europaplatz 3, 69115 Heidelberg, Germany

Printed by Libri Plureos GmbH
in Hamburg, Germany